W9-AWP-969

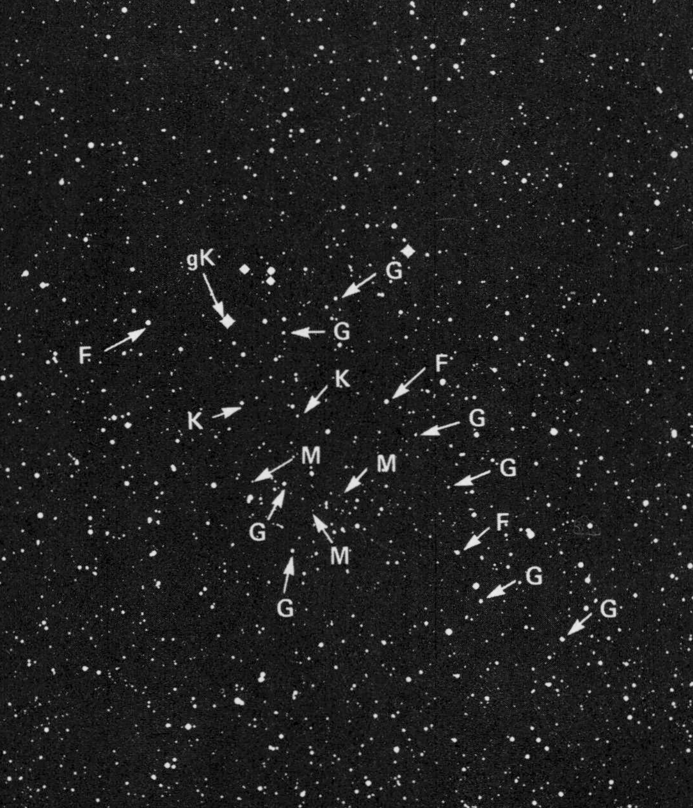
gK
G
G
F
F
K
K
G
M
M
G
G
F
M
G
G
G

THE SEARCH FOR EXTRATERRESTRIAL INTELLIGENCE

Prepared by
National Aeronautics
and Space Administration

Edited by
Philip Morrison,
John Billingham
and
John Wolfe

DOVER PUBLICATIONS, INC.
NEW YORK

Frontispiece: A photographic image of a star field with the addition of designations of stellar type for some of the stars. The letter designation (e.g., A) indicates the spectral type of the star. The conventional spectral types are O, B, A, F, G, K, and M, with O-stars being the hottest (effective surface temperatures in excess of 30,000 K) and M-stars being the coolest (effective surface temperatures of 3,000 to 4,000 K; the effective surface temperature of the Sun is about 5,800 K). The prefix "g" indicates that a star is a giant star, a star that has moved away from the main sequence. Stars indicated only with the spectral designation are main sequence stars, deriving their energy primarily from the conversion of hydrogen into helium. The two important aspects of the figure are first that stars like the Sun, spectral type G, are very numerous in the Galaxy, and second, in any typical group of stars, most stars are of spectral types G, K, and M. These types of stars are long lived (10 billion years or greater). The figure shows that viewed from the perspective afforded by interstellar distances, the Sun would be a rather common and ubiquitous type of object. This suggests that the Sun's retinue of planetary companions, and perhaps the intelligent life forms existing on one of these planets, may also be common and ubiquitous phenomena. (Illustration kindly provided by Prof. Jesse L. Greenstein, California Institute of Technology, Hale Observatory, Pasadena, California).

Published in Canada by General Publishing Company, Ltd., 30 Lesmill Road, Don Mills, Toronto, Ontario.

Published in the United Kingdom by Constable and Company, Ltd., 10 Orange Street, London WC2H 7EG.

This Dover edition, first published in 1979, is a republication of the work originally published in 1977 by the United States Government Scientific and Technical Information Office as NASA SP-419. In order to conserve paper, numerous part titles and blank pages in the government publication, as well as the record of Workshop Meetings held between January, 1975 and June, 1976, have been omitted from this edition, and the pages have consequently been renumbered. The main text is complete and unabridged.

International Standard Book Number: 0-486-23890-3
Library of Congress Catalog Card Number: 79-52011

Manufactured in the United States of America
Dover Publications, Inc.
180 Varick Street
New York, N.Y. 10014

FOREWORD

There are few questions that more excite the curiosity, the imagination and the exploratory bent of modern man than the one posed in this study: Are we humans alone in this vast universe? The question is usually expressed in terms of other possible intelligent beings, on other planets. The philosopher in me would want to believe that if there are other intelligent beings, they are also free, and will use that freedom to try to find us. The basic problem to which this study is addressed is similar: Will we use our freedom to find them? What priority should this search have for modern man, everywhere?

Few would disagree with the proposition that we are living in a truly revolutionary age, inaugurated by Sputnik and the first trip to the Moon. In another such age – the Copernican – the prevailing religious or theological thought resisted, with the then current wisdom, the proposition that the Sun and the Universe did not rotate around the Earth. They mistakenly believed that man was the center of the Universe, and that astronomy should reflect that anthropocentric belief. They may ultimately be right about man, though not about astronomy, if we do not ever find intelligent life elsewhere in the Universe. However, we should not be predisposed to accept the proposition that we indeed are alone and unique as creatures possessing intelligence and freedom in this whole vast Universe.

I must now mention God – otherwise quite properly unmentioned in these scientific studies – and must go a step further and pose the question: Can a religious person, or even more, a theologian, possibly be legitimately involved in, even be excited by these discussions of the possibility of other intelligent and free creatures out there?

Just last week, I was discussing the subject with a Russian lawyer who regarded me with some surprise and asked: "Surely you must abandon your theology when you consider these possibilities?" "Indeed, I don't," I replied. "It is precisely because I believe theologically that there is a being called God, and that He is infinite in intelligence, freedom, and power, that I cannot take it upon myself *to limit what He might have done*." Once he created the Big Bang – and there had to be something, call it energy, hydrogen, or whatever, to go bang – He could have envisioned it going in billions of directions as it evolved, including billions of life forms and billions of kinds of intelligent beings. I will go even further. There conceivably can be billions of universes created with other Big Bangs or different arrangements. Why limit Infinite Power or Energy which is a name of God? We should get some hint from the almost, but not quite, infinite profusion of the Universe we still know only in part. Only one consideration is important here regarding creation. Since God is intelligent, however He creates – "Let there be light" – Bang – or otherwise, whatever He creates is a cosmos and not a chaos since all His creation has to reflect Him. What reflects Him most is intelligence and freedom, not matter. "We are made in His image," why suppose that He did not create the most of what reflects Him the best. He certainly made a lot of matter. Why not more intelligence, more free beings, who alone can seek and know Him?

As a theologian, I would say that this proposed search for extraterrestrial intelligence (SETI) is also a search of knowing and understanding God through His works – especially those works that most reflect Him. Finding others than ourselves would mean knowing Him better.

Theodore M. Hesburgh, C.S.C.
President, University of Notre Dame

MEMBERS

SCIENCE WORKSHOPS ON INTERSTELLAR COMMUNICATION

Philip Morrison, Chairman	Massachusetts Institute of Technology
Ronald Bracewell	Stanford University
Harrison Brown	California Institute of Technology
A. G. W. Cameron	Harvard University
Frank Drake	Cornell University
Jesse Greenstein	California Institute of Technology
Fred Haddock	University of Michigan
George Herbig	University of California-Santa Cruz
Arthur Kantrowitz	AVCO Everett Research Laboratory
Kenneth Kellermann	NRAO, Green Bank
Joshua Lederberg	Stanford University
John Lewis	Massachusetts Institute of Technology
Bruce Murray	Jet Propulsion Laboratory
Bernard Oliver	Hewlett-Packard
Carl Sagan	Cornell University
Charles Townes	University of California-Berkeley

PREFACE

Over the past two decades there has developed an increasingly serious debate about the existence of extraterrestrial intelligent life. More recently, there have been significant deliberations about ways in which extraterrestrial intelligence might in fact be detected. In the past two years, a series of Science Workshops has examined both questions in more detail. The Workshop activities were part of a feasibility study on the Search for Extraterrestrial Intelligence (SETI) conducted by the NASA Ames Research Center.

The objectives of the Science Workshops, as agreed at the second meeting in April 1975, were: to examine systematically the validity of the fundamental criteria and axioms associated with a program to detect extraterrestrial intelligent life; to identify areas of research in the astronomical sciences, and in other fields, that would improve the confidence levels of current probability estimates relevant to SETI; to enumerate the reasons for undertaking a search, the values and risks of success, and the consequences of failure; to explore alternative methods of conducting a search; to select, in a systematic way, preferred approaches; to indicate the conceptual design of a minimum useful system as required to implement the preferred approaches; to delineate the new opportunities for astronomical research provided by the system and their implications for system design; to outline the scale and timing of the search and the resources required to carry it out; to examine the impact of conducting a search, and the impact of success or failure in terms of national, international, social and environmental considerations; and to recommend a course of action, including specific near-term activities.

This report presents the findings of the series of Workshops. The major conclusions of our deliberations are presented in Section I. First, an Introduction lays out the background and rationale for a SETI program, and then in The Impact of SETI, we examine the implications of the program. In particular, the Impact section examines the significance of the detection of signals and of information that may be contained in signals from extraterrestrial civilizations.

For those who wish to see some of the arguments in more detail, we have extracted from the discussions of the last two years, six of the most interesting and significant elements of the debate in the form of Colloquies (Section II). Finally, we have documented, in greater depth, a selection of detailed technical arguments about various aspects of the SETI endeavor. This is Section III – Complementary Documents.

The reader should note that the Introduction, the Impact of SETI, and the Conclusions, which comprise Section I of this volume, have been prepared by and represent the views of the Workshop as a whole. Sections II and III, on the other hand, have been prepared by the individual authors listed, and while consonant with the major SETI findings, reflect specifically the views and style of presentation of the authors.

In addition to the series of six Workshops, and at the instigation of the participants, two additional series of meetings were held. The first, under the Chairmanship of Dr. Joshua Lederberg of Stanford University, addressed the question of Cultural Evolution in the context of SETI. The second, under Dr. Jesse Greenstein of the California Institute of Technology, addressed the question of the Detection of Other Planetary Systems. The conclusions of these meetings are presented in Colloquies 2 and 3.

The last of the Complementary Documents (III-15) lists the members of the Science Workshops, our consultants and advisors, and the agendas for the nine Workshop meetings. Detailed minutes of all of the Workshops are available from Dr. John Billingham, SETI Program Office, NASA Ames Research Center, Moffett Field, California, 94035.

I would like to express my appreciation to everyone who has worked with me in this undertaking I must single out first the Workshop members themselves (see Complementary Document 15), and in particular Joshua Lederberg and Jesse Greenstein for their major contributions in taking the chair at their respective special Workshops (see Colloquies 2 and 3). The assistance of the NASA Centers, and specifically of the SETI Groups at the Ames Research Center and Jet Propulsion Laboratory must be recognized, together with numerous contributions from consultants and speakers who have addressed and advised us. Last, but by no means least, special thanks are due to Vera Buescher, Secretary to the Ames SETI Team, for her loyal and indefatigable attention to the thousand details which went into the preparation of this report.

In conclusion, I would hope that our report will provide a logical basis for the evolution of a thoroughgoing but measured endeavor that could become a significant milestone in the history of our civilization.

We recommend the initiation of a SETI program now.

Philip Morrison
Chairman

TABLE OF CONTENTS

III. COMPLEMENTARY DOCUMENTS

BRIEF TITLES FOR ILLUSTRATIONS

BRIEF TITLES FOR FIGURES

BRIEF TITLES FOR TABLES

View of Arecibo Observatory in Puerto Rico with its 300 m dish – the world's largest. A small fraction of its observation time is devoted to ETI searches.

SECTION I: CONSENSUS

INTRODUCTION

> *Heaven and earth are large, yet in the whole of space they are but as a small grain of rice . . . . It is as if the whole of empty space were a tree, and heaven and earth were one of its fruits. Empty space is like a kingdom, and heaven and earth no more than a single individual person in that kingdom. Upon one tree there are many fruits, and in one kingdom many people. How unreasonable it would be to suppose that besides the heaven and earth which we can see there are no other heavens and no other earths?*
>
> Teng Mu, 13th Century philosopher
> (translated by Joseph Needham)

In the enormous emptiness of space we can now recognize so many stars that we could count one hundred billion of them for each human being alive. Yet we know of only one inhabited planet, our Earth. The Earth has supported the development of life nurtured by one commonplace star, the nearby five-billion-year old Sun. We look out into the starry Universe quite unable to see within its compass any sign that we are not alone. The other planets near our Sun offer some hope to a search for other life, and indeed for many months Viking on the surface of Mars has been reporting the enigmatic chemical activity of the Martian soil. We remain uncertain, at the time of writing, whether the chemical changes are biological or inorganic in nature.

The web of life here on Earth is the consequence of a long complex sequence of natural selection by which life increased its scope and its variety, always exploiting the flood of energy bestowed directly or indirectly by the Sun. The Earth has seen fire and ice, yet it has provided steadily, for three billion years without a break, some environments to which life could adapt. Changes were never so drastic or so rapid that all survival became impossible, though particular species have arisen and died by the millions. Indeed, life has spread from its origins, probably near the seashore, to alpine peaks and ocean troughs, and has diversified almost beyond description. Our species and a few of our forebears have achieved considerable technological abilities and some degree of self-knowledge. Nor do we foresee any natural catastrophe ending this fabric of life until in due course the Sun itself runs out of nuclear fuel, some five billion years in the future.

We all know the starry sky at night, and on our deep photographs of the sky we see everywhere a dusting of small dots. Analysis of the light which caused those images, using its intensity and the details of its spectrum, has made it certain that such dots represent suns resembling our own, about which we know only that they are suns. Our own Sun with its cortege of planets would be just such another single dot, quite indistinguishable from a hundred million others at the distances we scan.

We have been able to understand in a general way how stars are born out of dense clouds of gas and dust in the interstellar spaces; we can see other stars in the transient stages of birth, as once was the Sun and its planets. Are planets always born in the spinning disk of gas out of which the other suns form? Or is our own set of planets as rare as its central luminary is commonplace? We cannot now say, though we are sure that the processes that form stars and might have formed planets as well, were going on for billions of years before our solar system formed, and will outlast our Sun.

If around those other visible suns there spin other planets, hidden from us by the distances of space, it is at least possible that on some the work of natural selection has continued for times which could be five or ten billion years longer than the whole history of our Sun and Earth. We could conceive that life never arose on a given planet, or that it exhausted its resources of adaptability, to end in an algal monotony, or in total extinction. Or we can imagine the slow evolution of beings – not human – able to control their world and themselves and to know the Universe.

In evolutionary diversity there is still unity. Squid and human see with eyes that evolved quite differently, and yet resemble each other, for they perform similar tasks. The big tuna, the extinct icthyosaur, and the dolphin resemble each other closely in streamlined form, and even somewhat in behavior, They are distinct evolutionary solutions to the problem of earning a living by predation upon fast-swimming fish; the three, fish, reptile, and mammal, have been molded alike by natural selection to solve the single dynamical problem of fast pursuit in the sea. Similarly, the way of life of *H. sapiens* appears to spread and to succeed; it seems to us that if natural selection has once built so subtle and successful a scheme, it can do so again. Sapient beings on other planets would in no way be our biological kin for they would share with us no common ancestor. But they might have converged with us in behavior; they might have evolved to culture, and then, say, to radio telescopes! Culture is a workable way of life, like hunting schools of mackerel. Indeed, we have seen that human cultural evolution, also, often converges: no less a development than writing was independently achieved by the Aztecs, the Chinese, and the peoples of the Middle East. On this basis, it would be consistent with the historical development of the great ideas of science to postulate that for a time of unknown duration, near an unknown number of stars, deliberate radio beacons or unintended radio leakage are present. This is a hypothesis untested, but capable of verification by experiment.

It is not idle curiosity to inquire whether other intelligent life has arisen and survived near some distant sun, beings in no way our biological kin. For by some sign of that presence we might come to learn, in a way, our own possible future. Indeed, the one most solid result of the calculus of chance which governs our thoughts about such uncertainty is this: intelligent beings out there – if they exist at all – almost surely form societies which have endured for a time long compared to the history of our own civilization, a time which might even reach the span of geological time itself.

Astronomers have real hope of detecting planets near other stars, especially relatively neighboring ones, by new optical or infrared measurements from ground or orbit. But detection of plant or animal life implies a landing such as we made on Mars, and this is well beyond our capabilities over interstellar distances.

If we are to learn about distant life, it must make itself perceptible. As far as we can see, only life that has followed our own evolution to the extent of being able to send *some* mark of its presence across space can be found. This must mean that intelligence develops naturally out of evolving life, that it can make signals capable of traversing space, and that, for some period of time at least, it wants to make its presence known (or at least does not conceal it!). If these conditions exist anywhere, we might hope to detect creatures far older and more capable than ourselves. Exploration would then cross a new frontier; the frontier of an intelligence biologically wholly unrelated to our own.

How would such signals be made? Might super-Viking probes cross space? Might light flashes like stellar lighthouses show an intelligent presence? Much speculation has considered the situation (some of the variety of different ideas are presented in Complementary Document 1). The key facts seem to be that radio waves cross space well, and that the radio engineer has found means to detect extremely weak signals with large dishes and extremely sensitive receivers. Violent events on every scale, from explosions in galaxies to electrical instabilities on the planet Jupiter, have been recorded by radio astronomers. None of these signals appear to bear the marks of any but an astronomical origin, so far. Interesting as these have been, it remains true that radio energy compared to visible light is scarce in the Galaxy. Within the scope of present knowledge in our own Galaxy, a certain well-defined radio waveband (from about one meter to one centimeter) is, for natural reasons, the quietest region over the whole span of electromagnetic waves (see Section II-4).

This fact lies behind a remarkable event in human history. Almost imperceptibly, without really intending it, within the past two or three decades we have entered a new communicative epoch Until that time, we could have made no sound, no pattern or mark, no explosive flash of light on our small planet that could be detected far out among the stars by any means we understand. Space is too deep, and the stars are rivals too brilliant, for any mere faint human glow to become visible far away. Even the whole amount of sunlight reflected from a planet, a light source thousands of times more powerful than all the energy now at human disposal – is still beyond our ability to pick out at the distance of a nearby star. But our radio technique, only a generation or so old, has now reached such maturity that a signal sent from an existing radio dish on Earth, with sending and receiving devices already at hand, could be detected with ease across the Galaxy by a similar dish, if only it is pointed in the *right* direction at the *right* time, tuned to the *right* frequency. Such a lucky observer – or one who is patiently and systematically searching – would see us as unique, distinguished among all the stars, a strange source of coherent radio emission unprecedented in the Galaxy.

Or are we without precedent? Are we the first and only?

Or are there in fact somewhere among the hundred billion stars of the Galaxy other such beams, perhaps so many of them that our civilization, like our Sun, is to be counted as but one member of a numerous natural class? For such a radio beam cannot come, we think, from any glowing sphere of gas or drifting beam of particles. It can come only from something like our own complex artificial apparatus, far different from any star or planet, smaller, newer, much more particular; something we would recognize as the product of other understanding and ingenious beings.

That is the topic of this technical report: the search for extraterrestrial intelligence, SETI.

We do not intend to send any signals out to add to those which have already gone out from our TV transmitters and our powerful radars. Rather, we want to listen, to search all the directions of space, the many channels of the radio (and other) domains, to seek possible signals. Perhaps it will be only an accidental signal, as we have made ourselves. That would be harder to find. Or perhaps there is a deliberate signal, a beacon for identification, or even a network of communication. There seems no way to know without trying the search. This is an exploration of a new kind, an exploration we think both as uncertain and as full of meaning as any that human beings have ever undertaken.

The search would be an expression of man's natural exploratory drive. The time is at hand when we can begin it in earnest. How far and hard we will need to look before we find a signal, or before we become at last convinced that our nature is rare in the Universe, we cannot now know.

THE IMPACT OF SETI

Whether the search for extraterrestrial intelligence succeeds or fails, its consequences will be extraordinary.

If we make a long dedicated search that fails, we will not have wasted our time. We will have developed important technology, with applications to many other aspects of our own civilization. We will surely have added greatly to our knowledge of the physical Universe. The global organization of a search for interstellar radio messages, quite apart from its outcome, can have a cohesive and constructive influence upon our view of the human condition. But above all, we will have strengthened belief in the near uniqueness of our species, our civilization and our planet. Lacking any detection, the conviction of our uniqueness would hardly ever reach certainty; it would form over a long time, less into sharp conclusions than into a kind of substructure of human thought, a ruling consensus of attitudes. If intelligent, technological life is rare or absent elsewhere, we will have learned how precious is our human culture, how unique our biological patrimony, painstakingly evolved over three or four thousand million years of tortuous evolutionary history. Even a growing possibility of such a finding will stress, as perhaps nothing else can, our lonely responsibilities to the human dangers of our time.

On the other hand, were we to locate but a single extraterrestrial signal, we would know immediately one great truth: that it is possible for a civilization to maintain an advanced technological state and *not* destroy itself. We might even learn that life and intelligence pervade the Universe. The sharpness of the impact of simple detection will depend on the circumstances of discovery. If we were to find real signals after only a few years of a modest search, there is little doubt the news would be sensational. If, on the other hand, signals were detected only after a protracted effort over generations with a large search system, the result might be less conspicuous.

Note well that it is likely that the early announcements of the detection of deliberate signals may turn out to be mistaken, not verified by further study and observation. They may be natural phenomena of a new kind, or some terrestrial signal, or even a hoax. (Indeed, this has already happened – more than once!) Press and public must use caution if we are to escape the volatile raising and dashing of great hopes. We stress the importance of a skeptical stance and *the need for verification*, because we hold that even a single genuine detection would in and of itself have enormous importance.

Of course it is very difficult to foresee the content of a signal except in the most general way. A signal could be a beacon – a deliberate transmission specifically for the purpose of attracting the attention of an emerging civilization like ourselves. Alternately, it could be a leakage signal similar to our own television broadcasts or radars, not intended for our detection. Whatever the signal, we would remind the reader that it will be a one-way transmission. Any messages in such a transmission would be a message between cultures, not between persons. We have human analogies at hand, in our long-continued interest in great books from the past, say the Greek philosophers; we ponder them afresh in each generation, without any hope of interrogating Socrates or arguing with Aristotle.

The first authentic signals will attract intense headline attention. But after that the pace must slow. Perhaps we will learn only that the signal exists. This alone will be significant. We will know we are not alone. However, the information content of any signal could be rich. Study would continue for decades, even generations. Books and universities will be more suited for the news than the daily programs. If the signal is deliberate, decoding will be relatively easy, we expect, because the signal will be anticryptographic; made to reveal its own language coding. If the message comes by radio, both transmitting and receiving civilizations will have in common at least the details of radiophysics. (The commonality of mathematics and the physical sciences is the reason that many scientists expect the messages from extraterrestrial civilizations to be decodable – if in a slow and halting manner.) No one is wise enough to predict in detail what the consequences of such a decoding will be, because no one is wise enough to understand beforehand what the nature of the message will be.

Some have worried that a message from an advanced society might make us lose faith in our own, might deprive us of the initiative to make new discoveries if it seems that there are others who have made those discoveries already, or might have other negative consequences. But we point out that we are free to ignore an interstellar message if we find it offensive. Few of us have rejected schools because teachers and textbooks exhibit learning of which we were so far ignorant. If we receive a message, we are under no obligation to reply.* If we do not choose to respond, there is no way for the transmitting civilization to determine that its message was received and understood on the tiny distant planet Earth. (Even a sweet siren song would be little risk, for we are bound by bonds of distance and time much more securely than was Ulysses tied to the mast.) The receipt and translation of a radio message from the depths of space seems to pose few dangers to mankind; instead it holds promise of philosophical and perhaps practical benefits for all of humanity.

Other imaginative and enthusiastic speculators foresee big technological gains, hints and leads of extraordinary value. They imagine too all sorts of scientific results, ranging from a valid picture of the past and the future of the Universe through theories of the fundamental particles to whole new biologies. Some conjecture that we might hear from near-immortals the views of distant and venerable thinkers on the deepest values of conscious beings and their societies! Perhaps we will forever become linked with a chain of rich cultures, a vast galactic network. Who can say?

If it is true that such signals might give us, so to speak, a view of one future for human history, they would take on even greater importance. Judging that importance lies quite outside the competence of the members of this committee, chosen mainly from natural scientists and engineers. We sought some advice from a group of persons trained in history and the evolution of culture, but it is plain that such broad issues of the human future go beyond what any small committee can usefully outline in a few days. The question deserves rather the serious and prolonged attention of many professionals from a wide range of disciplines – anthropologists, artists, lawyers, politicians, philosophers, theologians – even more than that, the concern of all thoughtful persons, whether specialists or not. We must, all of us, consider the outcome of the

---

*It is for this reason that this undertaking is not called *Communication* with Extraterrestrial Intelligence (CETI), but *Search* for Extraterrestrial Intelligence (SETI).

search. That search, we believe, is feasible; its outcome is truly important, either way. Dare we begin? For us who write here that question has step by step become instead: *Dare we delay?*

FIRST CONCLUSION

1. IT IS BOTH TIMELY AND FEASIBLE TO BEGIN A SERIOUS SEARCH FOR EXTRATERRESTRIAL INTELLIGENCE

Only a few decades ago most astronomers believed that planetary systems were extremely rare, that the solar system and the habitat for life that Earth provides might well be unique in the Galaxy. At the same time so little was known about the chemical basis for the origin of life that this event appeared to many to verge on the miraculous. No serious program for detecting extraterrestrial intelligence (ETI) could arise in such an intellectual climate. Since then numerous advances in a number of apparently diverse sciences have eroded the reasons for expecting planetary systems and biogenesis on suitable planets to be unlikely. Indeed, theory today suggests that planetary systems may be the rule around solar type stars, and that the Universe, far from being barren, may be teeming with life, much of it highly evolved. (See Section II-1 and II-3.)

During the latter half of the last and the first part of this century, the slow rotation of the Sun stood as a formidable objection to the nebular hypothesis of Kant and Laplace, which proposed that planetary systems formed out of the same condensing cloud that produced the primary star. An initial rotation rapid enough to produce the Sun's planets should have produced a Sun spinning a thousand times faster – too fast to become a spherical star. As a result, various "catastrophic" theories of the origin of the solar system were proposed, all of which depended on events so rare as to make the solar system virtually unique.

Then, in the late 1930's, Spitzer showed that starstuff torn out by tidal or concussive forces would explode into space rather than condense into planets. Shortly thereafter research into plasma physics, and observations of solar prominences, revealed the magnetohydrodynamic coupling of ionized matter to magnetic fields, a mechanism whereby stars in the process of formation can slow their rotation. As a result, the theory in which planets condense out of the whirling lens of gas and dust that will become a star has regained wide acceptance. Planetary systems are now believed to exist around a substantial fraction of stars. (See Section II-3.)

Meanwhile the discoveries that the organic building blocks for DNA and proteins can be formed by natural processes out of molecules comprising the early atmosphere of Earth, and that many organic molecules are even formed in the depths of interstellar space, have made the spontaneous origin of life on suitable planets seem far more probable. Life appears to have developed on Earth almost as soon as seas had formed and chemical evolution had provided the building blocks. Earth has been lifeless for only a small fraction of its age. This leads many exobiologists today to look upon life as a very likely development, given a suitable planet. (See Section II-1.)

The present climate of belief makes it timely to consider a search for extraterrestrial life, but is such a search feasible? It is certainly out of the question, at our present level of technology or, indeed, at any level we can foresee, to mount an interstellar search by spaceship. On the other hand, we believe it is feasible to begin a search for signals radiated by other civilizations having technologies at least as advanced as ours. We can expect, with considerable confidence, that such signals will consist of electromagnetic waves; no other known particle approaches the photon in ease of generation, direction and detection. None flies faster, none has less energy and is therefore cheaper than the radio frequency photon. It has long been argued that signals of extraterrestrial origin will be most apt to be detected in the so-called microwave window: wavelengths from about 0.5 to 30 cm. Natural noise sources rise to great height on either side of this window, making it the quietest part of the spectrum for everyone in the Galaxy. We concur with these arguments. (See Section II-4.)

Existing radio telescopes are capable of receiving signals from our interstellar neighbors, if of high power or if beamed at us by similar telescopes used as transmitters. The large antenna at Arecibo could detect its counterpart thousands of light years away. Indeed, it could detect transmissions from nearby stars less powerful but similar to our own television and radars.

Terrestrial UHF and microwave emanations now fill a sphere some twenty light years in radius. This unintended announcement of our technological prowess is growing stronger each year and is expanding into space at the speed of light. The same phenomenon may well denote the presence of any technological society. In fact, our own radar leakage may have already been detected by a nearby civilization. In addition, advanced societies may radiate beacons for a variety of reasons, possibly merely to bring emerging societies into contact with a long established intelligent community of advanced societies throughout the Galaxy. A search begun today could detect signals of either type.

We propose a search for signals in the microwave part of the radio spectrum, but not at this time the sending of signals. Even though we expect our society to continue to radiate TV and radar signals we do not propose to increase our detectability by, say, intentionally beaming signals at likely stars. There is an immediate payoff if we receive a signal; transmission requires that we wait out the round trip light time before we can hope for any results. Transmission should be considered only in response to a received signal or after a prolonged listening program has failed to detect any signals. (See Section II-5.)

Not only is the technology for discovering ETI already at hand, but every passing year will see the radio frequency interference (RFI) problem grow worse while only modest improvements in technology can occur. (See Sections III-8 and III-9.) Perfect receivers would not double the sensitivity of a search system over that which we can already achieve. Given optimum data processing, large increases in sensitivity are to be had only by increasing collecting area. It is true that data processing technology is improving rapidly, but presently achievable data processing technology is adequate and inexpensive. Further, the techniques need to be developed in association with existing facilities and comprehensive searches made before it becomes evident that a more sensitive system is needed. Great discoveries are often the result more of courage and determination than of the ultimate in equipment. The Niña, the Pinta, and the Santa Maria were not jet airliners, but they did the job.

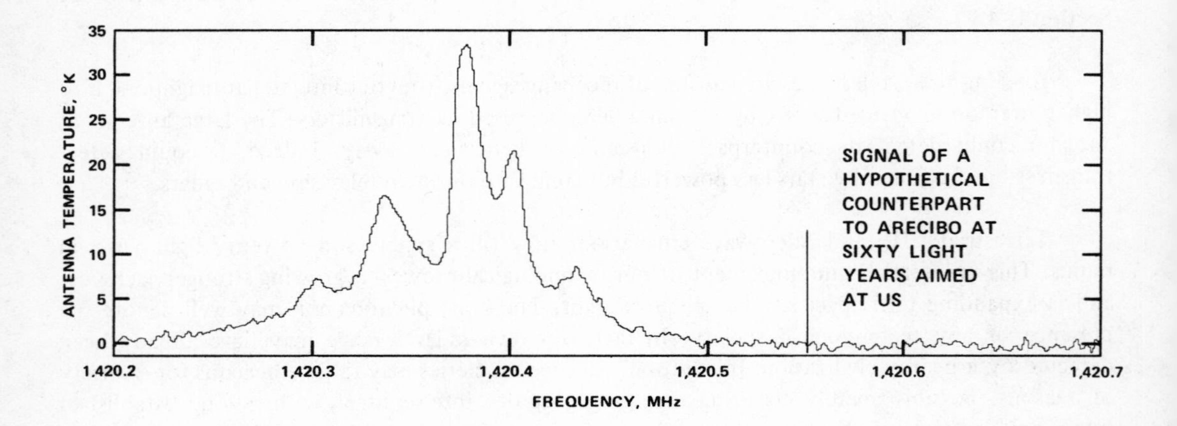

Portion of output scan from 1024 channel (of 1 KHz each) analyzer at Arecibo Observatory showing an ETI search, centered on HI line, with α-Ophiuchi as the target. Structure shown is due to interstellar hydrogen clouds at various drift velocities. Bar indicates strength of hypothetical signal that could be received from a transmitter with the EIRP of Arecibo, located at the 60 ly distance of α-Ophiuchi.

SECOND CONCLUSION

2. A SIGNIFICANT SETI PROGRAM WITH SUBSTANTIAL POTENTIAL SECONDARY BENEFITS CAN BE UNDERTAKEN WITH ONLY MODEST RESOURCES

A large, expensive system is not now needed for SETI. If we but equip existing radio telescopes with low-cost state-of-the-art receiving and data processing devices, we will have both the sensitivity to explore the vicinity of nearby stars for transmitters similar to Earth's, and to explore the entire Galaxy for more powerful signals, or for signals beamed at us. Such explorations, even should they yield negative results, would decrease our uncertainty concerning whether intelligent life transmitting powerful signals may lie beyond our solar system. At the very least, it would be of great interest and some importance either to know we have near neighbors, or to be reasonably confident no nearby transmitting civilizations exist. If, *after* we have made such modest searches, it seems important to us to embark upon a more ambitious SETI program, such as contemplated by the Cyclops study, the experience we will have gained will prove not only invaluable, but essential. Moreover, we expect to derive spin-off benefits of no small significance.

SETI Hardware

The arguments for electromagnetic waves as the communications medium seem compelling. The case for the microwave window seems very strong. The reasons for preferring the low end of the window are also strong, but not so strong that higher frequencies in the window should be ignored. The "water hole" between the H and OH lines is an especially attractive band that may be ideal for long range beacons. (See Sections II-4 and III-1.)

ETI signals, particularly those intended for detection by other searching societies, will probably be narrow in bandwidth compared with natural sources and may have monochromatic components which are as narrow as the interstellar medium permits. This increases their detectability at a given radiated power and distinguishes them from the natural background. The hardware needed for SETI therefore consists of an antenna or antenna system, low-noise wide-band receivers to cover the low-frequency end of the microwave window, means of resolving the received spectrum to a very high degree and means to search out and identify automatically any spectral anomalies.

Since halving the system noise temperature is equivalent to doubling the system sensitivity, it is important in SETI to have the lowest noise receivers that can be built. The background temperature in the preferred frequency region is only 6 K to 8 K (3 to 5 K in space) so every degree of reduction in receiver noise temperature is significant. The development of suitable low noise receivers represents a simple extension of present microwave technology and is not an expensive program. It would also benefit deep space communications and radio astronomy. (See Section III-5.)

To search for narrow band signals that may be anywhere in a wide frequency band and to do so in a reasonable time has been one of the major challenges of a SETI. In the Cyclops system concept the received signal was optically transformed into a high-resolution power spectrum. Since

1971 the growth of large-scale integrated circuit technology has been spectacular. It now appears possible to build, at reasonable cost, solid state fast Fourier analyzers capable of resolving the instantaneous bandwidth into at least a million channels on a real time basis. Development of such equipment is again a modest undertaking and the equipment would be very valuable for many other uses besides SETI. (For example see Section III-5.)

To complete the data processing it is necessary to examine the power spectrum or a succession of samples of the power spectrum for any sort of significant pattern such as a sustained peak that may drift slowly in frequency, a regularly recurring peak, or arrays of regularly spaced peaks, to name but a few. The data rates are so great that this pattern recognition must be automated. The principal problems associated with the pattern recognition system are the amount of data storage needed and the identification of the types of patterns to be recognized. Only a few years ago these could have presented severe difficulties, but the solid state electronics revolution has so reduced the cost of memory, that prospective data processing costs appear to be relatively inexpensive.

It has been estimated that the development of the right data processing equipment would increase the capability of existing radio telescopes to detect ETI signals by about a thousandfold. This means that very significant searches can be made using existing antennas so equipped and it is recommended that the search begin in this way. The possibility of discovering some unknown type of natural source in this way must not be overlooked.

Search Strategies

It is not feasible to search for all kinds of signals at all frequencies from all directions to the lowest flux levels at which a known signal of known frequency and direction of arrival can be detected. (See Sections II-5 and III-2.) The more inclusive the search becomes in frequency or spatial direction, the more time is required, unless we sacrifice sensitivity. This is, of course, the reason for making use of all available *a priori* information and guesses as to preferred frequencies and likely directions of arrival. Many ingenious arguments have been offered for special frequencies and directions or even times; all can be given some weight as the search proceeds. On the other hand, every reduction in some dimension of the search is based on an assumption that may be wrong.

The strategy of searching nearby F, G, and K main sequence stars at ever increasing range seems very natural; the only life we know lives on a planet around a G2 dwarf star. This strategy takes us only as far into space as necessary to discover our nearest radiative neighbors around such stars. On the other hand, only slightly older cultures may be capable of radiating much more powerful signals, or they may know that life is to be found only around a few stars of a certain spectral class and age and may beam signals at these. As is true for stars, the nearest transmitters may not be the brightest. The strongest signals may come from advanced societies at great distances, whose transmitters may not even be near any stars.

For these reasons it is premature to adopt only one strategy to the exclusion of others. To cover a wide range of other possibilities it is recommended that in addition to a high sensitivity search of nearby stars, there also be a complete search of the sky to as low a flux level and over as wide a frequency band as practicable. (See Sections II-5 and III-3.)

To be significant, a full sky survey should be able to detect coherent radiation at a flux level one or two orders of magnitude below that provided by existing radio astronomy surveys. This turns out to be easier than one might expect. Although a sky survey as sensitive as $\sim 3 \times 10^{-23}$ W/m^2 has been made this has covered only ~2% of the sky. Another, covering most of the sky, has been made to a sensitivity of $\sim 2 \times 10^{-20}$ W/m^2. But in these, as apparently in all radio astronomy sky surveys, any coherent signals that might have been present were rejected as "interference." Thus a complete sky survey using SETI data processing equipment to detect coherent signals at flux levels of $\sim 10^{-20}$ to $\sim 10^{-24}$ W/m^2 would be very significant. Existing antennas could be used to search the water hole to this level and the entire microwave window to as low as $\sim 10^{-23}$ W/m^2 in a few years of observing time.

The target search of the nearer F, G, and K main sequence stars should be conducted using SETI hardware with existing antennas. This would permit detection of coherent signals at a flux level as low as $\sim 10^{-27}$ W/m^2, or 10^3 to 10^7 times weaker than for the full sky search, assuming an observation time on the order of a half hour per star.

Both the sky survey and the targeted search could produce positive results, but even negative results will be of value since the upper limit flux levels that would result will be much lower than before. This could change our assessment of the capabilities of other intelligent life. The experience gained using SETI hardware in actual operation, with natural and man-made interference present, will affect the design of any future search strategies, and may lead to modifications of hardware, software, and search procedure. The searches we propose can be completed in approximately five years.

Planning a Dedicated Facility

SETI is more than a single effort. Like the exploration of the New World by our forefathers, like the present exploration of our solar system, it should be accomplished by many missions, each with some particular goal in mind. But there is a limit to the time that can be reasonably devoted to SETI from the facilities of radio astronomy or other services. To achieve the ultimate goals of SETI it will probably be important to have a dedicated SETI facility, the planning for which should begin now. This facility may never need to grow beyond a collecting area equivalent to one, or a few 100-m dishes. That will depend on future priorities, and on what we learn from the searches we immediately propose. The facility may be on the ground, or in space. (See Section III-7.) We should, however, keep possible future needs in mind, and be prepared to build it whenever and wherever it appears appropriate.

Supporting Activities

Several ancillary programs should be initiated and pursued. These include protection of the water hole (1.400 to 1.727 GHz) against radio frequency interference (RFI) (see Sections II-4, III-8, and III-9), the detection of extrasolar planetary systems (see Section III-3), the development of techniques for compiling extensive lists of target stars (see Section III-4), the study of alternative search strategies, and the continuing study of the cost effectiveness of space vs ground based systems.

In a resolution adopted at its fourth meeting the Science Workshop recommended that that international protection of the water hole against RFI be sought at the 1979 World Administrative Radio Conference. (See Section III-9.) Navigational satellite systems are presently being planned that would destroy the usefulness of this prime band of frequencies for SETI purposes. It is important to realize that for ground-based SETI systems such protection does not exclude all other services from the water hole, but only interfering ones such as satellites and nearby ground services. The RFI problem for space based SETI systems (especially systems in synchronous orbit) is more complex and probably more serious. Adequate shielding may be very expensive. It is not necessary that RFI protection of the water hole continue for all time. If no signals are found after a protracted sensitive search, the SETI priority may be relinquished.

The *sine qua non* of SETI is the plenitude of other planetary systems. While theoretical considerations suggest that planetary systems are common, it would be valuable to know *how* common and how their architecture varies with stellar class and multiplicity. Earlier astrometric telescopes and data reduction techniques could be improved to the point where the existence of near-by planets could be proved or disproved, but the effort might require two or three periods of a major planet, i.e., two or three decades. Preliminary calculations indicate that the direct observation of major planets around nearby stars should be possible with space telescopes of only modest size (on the order of one meter diameter). This could be accomplished by fitting the space telescope with a suitable filter or mask which greatly improves the contrast of a large planet with respect to the central star. Such an approach, if successful, would permit planets to be found in only two to three years after launch. This and other space techniques for direct planetary detection deserve active study and support. (See Section III-3.)

Present star catalogues list the coordinates of F, G, and K main sequence stars within only a few tens of light years of the Sun. If we ultimately carry on a search out to several hundred light years we will need to know the location of a thousand times as many target stars as are now listed. The problem of how best to conduct a whole sky star classification and cataloging program needs to be studied and, when solved, to be implemented. Since the compilation of such a target star data base must precede a major search, it is timely to begin the design study now. Both a greatly expanded catalogue of the solar neighborhood and knowledge about nearby planetary systems would be significant contributions to galactic and stellar astronomy as well as to SETI. (See Sections II-6, III-4, and III-6.)

Although it is assumed that the searches performed in this program will be mainly for narrow band signals at the low end of the microwave window, other possibilities should not be ignored. Given a matched filter a series of pulses is just as easy to detect as a continuous wave (CW) signal of the same average power. The pulsed signal, however, introduces the new dimensions of pulse shape, repetition rate, and duty cycle. At this same time it is not clear that CW signals are more probable than pulses. Continuing study of these and other alternatives is indicated.

It will be seen that the program advocated above is of modest scale yet has potential for both SETI success and scientific contribution. Above all it serves as a logical introduction to the future but does not constitute a blank check commitment to a large expensive effort. The program is not a dead end nor is it open ended. It will be timely to consider whether to proceed with a larger scale program after this earlier effort has shown us more accurately what might be involved.

THIRD CONCLUSION

3. LARGE SYSTEMS OF GREAT CAPABILITY CAN BE BUILT IF NEEDED

Large systems, involving construction of new antennas, are not now needed for SETI. Until we have completed an observational program as suggested in the Second Conclusion, there seems to be no reason to construct any facility much larger in scale than Arecibo. However, we may some day decide to embark on a more comprehensive search. This could require a system of great capability. Although we emphasize that we do not now recommend construction of such a system, we also feel that is is important to emphasize that a large SETI system is well within the capability of present-day technology.

The first feasibility study of large SETI systems was the 1971 Cyclops project. It concluded that we indeed have the technology to construct a very large ground-based phased array. The system considered would be capable of operating over the 1 to 10 GHz region of the microwave window, and could grow to collecting areas of many square kilometers if necessary. Its receiving system would be coupled to a data processing system capable of resolving 200 MHz of spectrum into 0.1 Hz channels and of detecting any coherent signal whose power equalled the noise power in this 0.1 Hz bandwidth.

At the request of the Ames Interstellar Communication Study Group, the Jet Propulsion Laboratory performed a detailed independent review of the Project Cyclops report, and found the study to be correct in its major technological conclusions. Today the data reduction would probably use large scale integrated circuit hardware exclusively, rather than optical processors. Today the system noise temperature could be nearer 10 K than 20 K. But these improvements only reinforce the basic conclusion that ground-based systems can be built that will detect a gigawatt *omnidirectional* beacon or its equivalent at a distance of 1000 light years. This corresponds, in the water hole, to a flux of one photon per second per square kilometer.

The principal cost of the Cyclops system was found to be the antennas. If the effectiveness of the data processing could be improved enough to double the sensitivity for the same antenna area, the original system performance could be achieved at about half the cost. Clearly in systems having large collecting area it is very important to make optimum use of that area by doing the best possible job of data processing. Further studies of the coherent signal detection problem and the possible tradeoffs in time and money vs antenna area are needed and should be started now.

Ground-Based vs Space Systems

Following the Cyclops study the Interstellar Communication Study Group at Ames contracted with the Stanford Research Institute to study various alternatives to a ground-based array in achieving large collecting area. A dozen alternatives were considered, four ground-based, four lunar-based, and four in space. (See Section III-7.)

The study revealed that very large, very lightweight, single unit antennas in space *may* be cost competitive with a large ground-based array. This conclusion can only be stated as a possibility and not as a fact because of the obvious difficulty of making valid cost comparisons between the well understood, mature ground-based antenna technology and the poorly understood, untested technology of large, lightweight space antennas.

In addition to the primary feasibility and cost of the space structures many other problems associated with space systems need further study. These include the means to shield the receiver against the severe RFI expected in space, the provision for wide band data links on a continuous basis, and the logistics of servicing and maintaining and operating a complex system in space. However, space systems also give unique advantage with respect to system noise, sky and frequency coverage and tracking ability. All that can be said at present is that space systems must be carefully considered in future plans.

Obviously, the whole question of Earth versus space based systems needs an order of magnitude more study before the issue can be resolved; this must be done before a commitment is made to any large search system. The possibility exists that a combination of ground and space systems would offer advantages not to be found in either alone.

Intermediate Steps

As discussed in the second conclusion, a small dedicated facility for SETI will probably eventually be desirable. This will most likely be a single new ground-based, or small space-based, antenna of advanced design, or both. If the facility is ground-based, it would be prudent if its site and design are chosen to ensure that the system be expandable at least to an intermediate size, such as a small array of 6 to 18 antennas. Such a system would increase the sensitivity well beyond that achievable with any existing antenna and would permit simultaneous searches using different strategies. It would also allow phasing techniques to be tested.

With respect to space borne antennas, it may be desirable, as studies proceed, to fly one or more medium size designs as shuttle payloads. The missions should be designed not only to test the structures but also to allow actual SETI and radio astronomy observations to be made in space. These antennas in conjunction with a dedicated ground facility could be used together as a very long baseline interferometer of greater capability than any now employed in radio astronomy. In addition observations could be made throughout the wide frequency bands over which the atmosphere is noisy or opaque.

Scientific Applications

As soon as a dedicated SETI facility achieves either a sensitivity or spectral coverage not found in present radio or radar astronomy instruments, it becomes a uniquely useful tool for research in these areas. An almost continuously increasing spectrum of applications exists as the SETI facility is expanded in scope. It is recommended that a fraction of the time of any dedicated facility be devoted to scientific research which that facility alone makes possible. This might well provide a series of discoveries which in themselves help justify the cost of the SETI facility. (See Sections III-5 and III-6).

We see that either in space or on the ground the SETI effort can efficiently grow from the initial effort to one using a very large system at whatever rate is appropriate. Early studies are needed to refine concepts of large systems, and especially to evaluate the usefulness of space. Even in the absence of the discovery of ETI signals, useful discoveries in science will accrue as the facility expands.

FOURTH CONCLUSION

4. SETI IS INTRINSICALLY AN INTERNATIONAL ENDEAVOR IN WHICH THE UNITED STATES CAN TAKE A LEAD

The Search for Extraterrestrial Intelligence offers benefits for all nations. The search would certainly be facilitated by, and may even require, international cooperation. It is a serious exploration, as important as any ever undertaken, and surely of larger scope than the journeys to the Earth's poles early in the century. We can hope for relatively quick results, but must prudently prepare for a protracted effort. The program must be kept open and public in the spirit of international science and exploration. We can and should expect growing cooperation with investigators from many countries, both those already displaying interest and activity, as the Soviet Union and Canada, and others whose interest would grow.

SETI is not only a response to the spirit of exploration but is natural to the metaphysical view of modern man. The question "Are we alone?" is pertinent to the entire species, both to us and our descendants.

International cooperation is essential to solving the radio frequency interference problem discussed above. Furthermore, it is possible that antennas may be required at various places throughout the world or in space: a system beyond the borders of any single nation. It seems clear to us that the SETI effort should be cast as a cooperative international endeavor at the start and that appropriate international relationships should be established through existing or novel international organizational arrangements. Joint funding is a desirable goal for such an approach. In any case the extended period which may be required for the detection of extraterrestrial intelligence – much less communication – emphasizes the need for organizational and cultural support more enduring than typically characteristic of national programs.

There may be a particular opportunity for joint Soviet and U.S. efforts in SETI. The Soviets have already begun a preliminary search. Their published discussion of this problem indicates that considerable interest exists within the scientific community there (see Section III-11). The USSR is capable of substantial space technology should that prove important in the future. Finally, joint leadership of an international SETI program by the U.S. and the Soviet Union might constitute a logical continuation of the cooperative endeavors in space initiated by the Keldysh-Low agreements most recently responsible for the Apollo-Soyuz Test Program.

West European nations, especially West Germany, Holland, and England, have also evidenced increasing interest in new radio astronomical endeavors. Thus, the possibility of initiating a SETI program through bilateral or multilateral arrangements involving the U.S. warrant consideration as well.

The United States Can Lead in the SETI Endeavor

The United States has frequently demonstrated the will and foresight to take the initiative in programs of worldwide benefit. The U.S. space program has provided not only excitement and scientific knowledge, but numerous practical satellite services not for this country alone, but for the whole world. It is in this same spirit of providing a focal point for international cooperation and support that we feel the U.S. can and should take the initiative in SETI.

The material, technological and intellectual resources of the U.S. are such that a large-scale SETI program could be carried on indefinitely by this country alone without appreciable drain on the economy. There are good reasons for believing the net effect on the economy could be positive. Even if international cooperation and support were slow to materialize, we believe SETI remains a feasible and worthwhile U.S. endeavor.

The psychology of and mechanisms for international cooperation suggest that an international SETI effort is unlikely until one big nation, such as the U.S., seizes the initiative and invites serious participation by others. It is in this sense of initiative and not in the pursuit of narrow national advantage that we recommend a leading role for the U.S. in SETI.

Initiating the SETI Effort

To carry on a significant United States SETI effort, public funds must be committed explicitly, with the approval of both the legislative and executive branches of the Federal Government. The evolution of an appropriate federal program lies with Congress and the President, but can only follow much preparatory work supported by one or more existing agencies.

We recognize that successful administration of the SETI program will require leadership by an agency with:

a. a mandate to carry out scientific research and exploration, possibly requiring operations in space;

b. large scale project management experience;

c. the ability to successfully involve the U.S. and foreign scientific community in a large scale enterprise;

d. in-house expertise in the relevant fields of technology; and

e. long range goals compatible with SETI.

Since NASA clearly meets these criteria it is particularly appropriate for NASA to take the lead in the early activities of a SETI program. SETI is an exploration of the Cosmos, clearly within the intent of legislation that established NASA in 1958. SETI overlaps and is synergistic with long term NASA programs in space astronomy, exobiology, deep space communications and planetary science. NASA is qualified technically, administratively, and practically to develop a national SETI strategy based on thoughtful interaction both with the scientific community and beyond to broader constituencies.

We therefore recommend that NASA continue its pioneering initiative in studying and planning near-term activities in support of SETI, and we urge that NASA, in cooperation with other agencies, begin the implementation of SETI.

SECTION II: COLLOQUIES

(These colloquies are general discussions on matters of central importance to SETI. Much of the discussion has been abstracted from the minutes of the meetings of the Science Workshops.)

1. COSMIC EVOLUTION

Through the centuries, man has continually searched the sky for clues to his destiny. His imagination has been captivated by the stars, his mind challenged by the mystery of their origin and extent, and his spirit imbued with a thirst for some understanding of his role in the cosmos.

Scientific discoveries in fields as diverse as astronomy and molecular biology have brought us, in the course of only the last 15 years, closer to solving three timeless riddles which many cultures have attempted to explain: How did the Universe begin and develop? How did life originate and evolve? What is our place and destiny in the Universe?

This burst of interdisciplinary discoveries has given rise to new concepts of the origin of life from inanimate material on the primitive Earth, of the formation of planets and stars, of the synthesis of fundamental particles of matter, and of the beginnings of the Universe itself: all seem to be founded on the same basic laws of chemistry and physics. The conclusion that the origin and evolution of life is inextricably interwoven with the origin and evolution of the cosmos seems inescapable. Taken in its totality, this pathway, from fundamental particles to advanced civilizations, forms the essence of the concept of cosmic evolution.

To be sure, the sequence from primordial fireball to matter to stars to planets to prebiotic chemistry to life and to intelligence, is incomplete and even controversial in some of its details. However, a broad picture is emerging, a picture that is both imaginative and illuminating.

The Universe appears to have begun as an awesome primordial fireball of pure radiation, commonly referred to as the "big bang," some 15 billion years ago. The totality of matter in the Universe, probably in the form of the most fundamental particles in nature, namely electrons, protons and neutrons, was flung apart with tremendous speed. As the fireball expanded and cooled, thermonuclear reactions produced helium nuclei. Still further expansion dropped the temperature to the point where hydrogen and helium atoms formed by combination of the electrons with the protons and helium nuclei, but elements heavier than helium were not produced in appreciable quantities. During later phases of the expansion, gravitational forces probably acted to enhance any nonuniformities of density that may have existed, and thus began the hierarchy of condensations that resulted in galaxies, stars, and, ultimately, planets.

Galaxies apparently had their origin in roughly spherical, slowly rotating, pregalactic clouds of hydrogen and helium which collapsed under their own gravity. When the contraction had proceeded sufficiently, stars began to form, and their rate of formation increased rapidly. The observed distribution of stellar populations agrees qualitatively with this general picture of contraction of the evolving galactic gas cloud.

Stars, like living organisms, are not immutable. They are born, evolve, and die. A star begins as a globular fraction of a larger gas cloud. The globule contracts under its own gravity, compressing and heating the gas to incandescence. The luminosity steadily increases as gravitational potential energy is converted into heat. When the internal temperature becomes high enough to initiate nuclear reactions, the contraction stops: the star has entered the main sequence. When all its fuel is burned the star dies as a white dwarf or explodes as a supernova, depending on its mass.

Although hydrogen and helium were created in the big bang, the rest of the elements were formed inside stars and in stellar explosions. Thus, every one of the heavier atoms in our bodies, including the oxygen that we breathe, the carbon and nitrogen in our tissues, the calcium in our bones, and the iron in our blood, was formed through the fusion of lighter atoms either at the center of a star or during the explosion of a star.

Stars spend 99 percent of their active lives burning hydrogen. When the hydrogen in the core is converted to helium, contraction begins. Core contraction releases enough energy to ignite helium which then burns to carbon. While all stars may eventually contribute to the production of carbon, nitrogen, and oxygen, and other elements up to and including iron, the heavier elements are probably made by neutron capture and beta decay inside massive stars during their final stages of evolution or when these stars explode as supernovae. Supernovae, then, may be the primary means by which the elements which are created in stars are recycled back into the interstellar medium. Out of the material sprayed around the galaxy by these explosions, somewhere else a new star with rocky planets can form.

The Milky Way galaxy is one of some 10 billion galaxies in the presently observable Universe, and our Sun is one of some 300 billion stars in the galaxy. The striking fact is that our galaxy and our Sun, to which we owe our very existence, are not unusual in any fundamental way compared with other galaxies and stars. Astronomical data indicate no peculiarities in origin, location, mass, luminosity, age, or other characteristics. Our Sun, therefore, is a relatively common type G dwarf star situated in a typical part of the disk of a rather typical spiral galaxy.

Discoveries made during the last 30 years have resulted in increased support for the nebular theory of the formation of planetary systems. According to this theory, development of the solar nebula was the result of the collapse of a large, rotating interstellar cloud whose inner portion was then heated by absorption of infrared radiation emanating from the proto-sun. At this stage, the cloud would be greatly flattened by its rotation and would consist of a disk with a denser, rapidly rotating, hot central region. Irregular density distributions in the disk are believed to have produced nucleation centers for the accretion of material into planets. In the outer regions of the disk most of the primordial gases of the disk probably went into forming the dense atmospheres of Jupiter-type planets. In the inner part of the disk, terrestrial-type planets may have formed early atmospheres which were then dissipated by the central star and replaced by outgassing from the planetary interiors; or they may have formed without atmospheres, the hydrogen and helium having been blown out of the inner solar nebula by the young Sun.

This renewed scientific support for the nebular theory of planetary formation has rekindled interest in the existence of extraterrestrial life because it predicts that stars with planetary systems should be the rule rather than the exception. In contrast to other theories, which attribute planetary formation to catastrophic events like star explosions or collisions, the nebular theory suggests that formation of planets will usually accompany the formation of a star. This implies that the galaxy and Universe should be replete with potentially life-supporting planetary sites.

Planets around other stars have never been observed directly with telescopes because of the great brightness difference between the star and planet, and the proximity of their images. Indirect evidence for the existence of another planetary system has been obtained by observation of the motion of Barnard's star over a 30-year period, but is inconclusive. The oscillation in its movement which, if real, could be accounted for by the presence of a planet of roughly Jupiter's size may be only instrumental error. At the present time, techniques are being developed which, in the near future, may allow direct observation of extra-solar planets (see Section II-3).

If the planets did condense from the proto-solar disk, their initial composition might be expected to reflect that of the Sun. The atomic abundances of some of the elements found in the Sun are, in order: hydrogen (87.0 percent), helium (12.9 percent), oxygen (0.025 percent), and nitrogen (0.02 percent). It should be noted that, with the exception of helium, these are the very elements which constitute 99 percent of living matter.

The giant planets appear to have a similar composition; that is, less than 1 percent heavy elements and the remainder hydrogen and helium. Furthermore, Jupiter contains methane, water, and ammonia, the reduced forms of carbon, and oxygen and nitrogen, as would be expected based on thermodynamic properties of these elements in the presence of a large excess of hydrogen. Hence, the present Jovian atmosphere may be in fact its primitive atmosphere, retained for billions of years because of the planet's mass. The inner planets, on the other hand, are composed almost entirely of heavy elements. With their present masses, these planets cannot prevent hydrogen or helium from escaping. Hence, the Earth's primitive atmosphere may have been transitory, with a more stable atmosphere arising as a result of outgassing of the crust by volcanic action. Volcanos discharge large quantities of water vapor and carbon dioxide, some nitrogen, and other gases. The carbon dioxide and water could be reduced to hydrogen and methane by free iron present in the early crust, and hydrogen and nitrogen could combine to form ammonia. Therefore, the Earth's early atmosphere may have consisted mainly of hydrogen, nitrogen, methane, ammonia, water, and carbon dioxide, with small amounts of other gases. It is believed that chemical evolution leading to life on Earth began in such an environment some 4.5 billion years ago.

The theory of chemical evolution was contemplated by Darwin, conceptualized by Oparin and Haldane, and tested experimentally by Miller and Urey. Simply put, the theory proposes that release of energy in the Earth's primitive atmosphere by various mechanisms resulted in the synthesis of simple organic molecules, which in turn were converted into molecules of greater complexity. New chemical reactions gradually came into being as a result of the increasing complexity, and new chemical order was imposed on the more simple organic chemical relations. At some point in this process, the first self-reproducing molecules appeared.

The energy sources available for the synthesis of organic compounds under prebiotic conditions were ultraviolet light from the Sun, electric discharges, ionizing radiation, and heat. Most of these have now been used in laboratory simulations to produce a wide variety of organic molecules from the presumed primitive atmosphere of methane, ammonia, water, and hydrogen.

These simulation reactions yield some compounds identical with those found in the complex biochemical structures of present-day organisms.

Among the compounds synthesized in these experiments are amino acids (the precursors to the proteins of living systems), purines and pyrimidines (monomer units of the nucleic acid genetic material), carbohydrates, hydrocarbons, fatty acids, and other compounds of structural and metabolic significance. Further prebiotic simulations have produced compounds with striking properties. For example, amino acids have been condensed to form proto-proteins which display low levels of enzymatic activity. Also, it has been shown that organic compounds formed under prebiotic conditions can aggregate into more complex structures that display chemical, structural, and physical properties remarkably similar to those of living cells. These discoveries are exciting because they provide models for the microenvironments in which specific compounds could be concentrated and reactions of importance to biological systems could more readily occur.

Besides these simulation experiments, two other lines of evidence support the theory of chemical evolution, and indicate that such syntheses are indeed universal. Recent radioastronomical observations have detected the presence, in the inhospitable environment of interstellar space, of ammonia, water vapor, formaldehyde, carbon monoxide, hydrogen cyanide, cyanoacetylene, acetaldehyde, formic acid, methanol, and a host of other compounds that are known precursors or intermediates in the chemical evolution simulation experiments. There is even a suggestion of highly polymeric material, like porphyrins or polyaromatic hydrocarbons. Cometary spectra indicate that a variety of organic compounds may be present. Since comets are considered to be similar in composition to the primordial material of the solar nebula, this constitutes evidence of organic matter in the very material from which the solar system was formed.

The analysis of meteorites has very clearly indicated that organic matter is present, with carbonaceous chondrites containing as much as 5 percent. Sophisticated organic chemical analyses have identified amino acids and other biologically significant compounds in meteorite samples. The nature of the compounds, their optical properties, and the distribution of isotopes within the molecules all argue conclusively for their being indigenous to the meteorite and therefore of extraterrestrial, abiological origin. Organic matter appears to be common in the cosmos.

To this point, then, the theory of chemical evolution is reasonable, understandable, and well-supported by experimental evidence. However, the sequence of events between the time when only a mixture of organic precursors existed in the primitive seas of the Earth and the time when, according to the geological record, the first living cell appeared some 3 billion years ago, is still a mystery. It is the only portion of the entire chain of events, culminating in man, for which substantive theories and data are lacking. And it is the crucial step for it marks the transition from nonliving to living systems. Somehow, organic molecules in the primitive ocean were assembled into that complex unit of life, the cell. The prebiotic simulation experiments and the terrestrial fossil record do, however, provide one significant inference: Processes leading from organic chemicals to living systems may take place over a relatively short period of time in the lifetime of a planet.

Over the next 3 billion years, the primitive organisms on Earth slowly evolved into the vast array of living systems we see today. Two cornerstones in the evolution of life were the development of photosynthetic capability, which is thought to have resulted in conversion of the atmosphere to its present oxidized state and which permitted cells to derive a great deal more energy from nutrient molecules, and sexual reproduction, which allowed advantageous mutations to be combined in a single individual.

The basic mechanism underlying biological evolution is mutation, the modification of the structure of the genetic material, and the retention in the gene pool of favorable traits. A favorable mutation confers a greater chance of survival since the cell or organism can now compete more successfully for energy sources and can better withstand environmental stresses. Over many generations the organisms possessing favorable mutations will gradually displace those without them. This is the essence of natural selection, originally proposed by Darwin and Wallace as a rational explanation for the whole history of the evolution of widely differing species that make up the plant and animal kingdoms. Experimental evidence from genetic research, the fossil record, and comparative biochemistry of present-day species supports the theory so completely that few have any major reservations as to its validity.

Man appeared very late in this sequence of events and, with his increased intelligence, came civilization, science and technology. Cultural evolution began and has proceeded very rapidly in the last few millenia (see Section II-2). An infinitesimal fraction of the matter of the Universe has been converted into the organic matter of the human brain. As a result, one part of the Universe can now reflect upon the whole process of cosmic evolution leading to the existence of human thought. We wonder whether this process is a frequent occurrence in the universe: in so doing we come to the postulate that life is widespread in the Universe, and at least in some cases, that this life may have evolved to the stage of intelligence and technological civilizations that it did on Earth. Some of these civilizations may be much more advanced than ourselves. They may have learned to communicate with each other, and achieved major advances in their own evolution as a result. Can *we* detect *them*?

Although many gaps, puzzles, and uncertainties still remain, this unifying concept, in which the expansion of the Universe, the birth and death of galaxies and stars, the formation of planets, the origins of life, and the ascent of humans are all explained by different features of the process of cosmic evolution, provides a sound scientific rationale on which to base a program to search for extraterrestrial intelligence (see Introduction and Conclusion 1 of Section I).

Prepared by: Ichtiaque S. Rasool
Office of Space Sciences
NASA Headquarters

Donald L. DeVincenzi
Planetary Biology Division
Ames Research Center

John Billingham
SETI Program Office
Ames Research Center

2. CULTURAL EVOLUTION

How likely is the evolution of intelligent life and technological civilization? Assuming that some self-replicating molecule that we can call "alive" evolves on some planet, will biological evolution act over the eons to produce a diversity of living forms similar to that we see on Earth? Might many of those forms develop intelligence and technology, or is man but a fluke that arose on Earth only as a result of a combination of highly improbable circumstances? What factors determine whether or not biological evolution will lead to intelligent and technological species?

These questions have been asked many times, but satisfactory answers have not been forthcoming. Perhaps this is because we have lacked the data to enable us to understand evolution to the necessary degree. The past two decades have seen a revolution in our understanding of animal behavior and a deluge of paleontological discoveries, especially of fossil hominids. As a result, we are, perhaps for the first time, able to make a rough assessment of the probability that evolution will produce an intelligent, technological species, and have some degree of confidence that the assessment is correct. Our new knowledge has changed the attitude of many specialists about the generality of cultural evolution from one of skepticism to a belief that it is a natural consequence of evolution under many environmental circumstances, given enough time.

On November 24 and 25, 1975, NASA sponsored a Workshop[1] on the evolution of intelligent life and technological civilizations which brought together prominent physical, biological, and social scientists (see Section III-15). That workshop endorsed the estimate that the combined probability that both intelligent life and technological civilization would evolve, assuming the origin of life on an arbitrary planet, is at least 10^{-2}. The workshop considered the opinion of George Gaylord Simpson, who was not present, that the evolution of intelligence was so improbable that perhaps it did not even happen on Earth. Albert Ammerman, geneticist from Stanford, pointed out that when Simpson formed his opinion, in the early 1950s, much less was known about such things as the evolution of the hominids, the social structure and communicative abilities of the chimpanzees, and the social behavior of other animals. At that time man appeared terribly unique; today this is no longer true. It is easy to imagine how other lines could have led or might yet lead to intelligence and civilization here on Earth. Intelligence, complex social organizations, and even the manipulative ability which makes possible the use of tools, have been demonstrated to have positive survival value; hence evolutionary pressures will tend to produce them.

A recurrent theme of the workshop was that it is wrong to focus on particular "crucial" events in the long evolutionary chain that has led to our modern technology. That these events appear by themselves to be improbable is misleading, since a low probability for a particular event does not imply a low combined probability for all possible events similar to it. Furthermore, that there can be a high probability of certain results, however improbable each of the innumerable ways they may come about, has been demonstrated many times by biological evolution itself.

---

[1]A Workshop on Cultural Evolution. Chairman, Dr. Joshua Lederberg. Held at the Center for Advanced Study in the Behavioral Sciences, Palo Alto, Calif.

Hans Lukas Teuber, neurophysiologist from MIT, pointed out that many organs have polyphyletic origins. The eye has been invented at least three times. The cephalopod eye, the insect eye, and the vertebrate eye all have totally different, independent evolutionary histories, but these histories have converged to essentially the same result. The neural networks in each of these eyes are shockingly similar. Indeed, any life-form evolving in an environment where the optical spectral band is important might well develop a light sensing organ with similar nerve structure. Nevertheless, octopi and vertebrates cannot see the same things. Octopi cannot distinguish mirror images. Therefore different evolutionary pathways cannot be expected to produce identical results. It is not unlikely that technological species are abundant in our galaxy, but it is very unlikely that elsewhere we will find men.

There was a wide range of opinions on what evolutionary factors were responsible specifically for hominid intelligence; probably many were important. Joshua Lederberg, geneticist from Stanford, thought that intraspecific warfare played a significant role. War seems to require rapid invention. Strategy discussions that are connected with the planning of warfare tend to involve a kind of verbal competition that is highly inventive. Furthermore, intraspecific conflict makes special demands on organisms that their battle with the environment does not: the difference is between intelligence against intelligence on the one hand, and intelligence against mere non-intelligence on the other. Finally, warfare seems to involve the young; organisms not suited to it suffer the consequence that their genes are eliminated from the breeding population. However, the necessity of the evolution of such an institution as warfare cannot be said to be established. Territoriality is a common trait among Terran animals, and some of the social mammals, such as man and the hyena, exhibit, as one form of territorial behavior, organized violent conflict between social groups. But territoriality is not a basic biological trait; many species do not exhibit it. Bernard Campbell, anthropologist from UCLA, thought it was but a result of the primary need to compete for resources, and that other means of allocating resources might be possible.

Probably the most important stimuli to the development of intelligence in early hominids was the demands of communication and language. J. Desmond Clark, paleoarcheologist from the University of California at Berkeley, pointed out that an increased rate of evolution of the brain set in about 3 million years ago, and this was correlated with increased use of stone tools. Manufacture of artifacts is evidence of complex social structure which in turn implies need for improved communication; at the very least the techniques of manufacture must be taught to the young. But John Eisenberg, ethologist from the Smithsonian Institute, thought that some of the hominids' increased cranial capacity was related to a general increase in motor coordination. The "text book idea" that man is a puny beast is simply not true. He is a fantastically powerful and coordinated organism, especially in the hands and limbs. He has subtle and accurate motor control which gives him great physical ability. He has independent control of his fingers and motor control of vocalization. And he has a very complex feedback system which enables him accurately to determine the course of thrown projectiles, with a little practice.

There was general agreement that the need to adapt to a predatory way of life on the savannah stimulated at least the early development of manipulative ability, motor coordination, and complex social organization in the hominids. The arboreal environment of the hominids' ancestors cannot have produced these traits. No monkey or ape can control a thrown projectile the way a man can; independent finger control is a uniquely hominid characteristic. Moreover, chimpanzees and other apes, though they use natural objects such as sticks for tools, have never developed a systematic tool-making ability. In captivity they can be taught to chip flakes from stone, but that they do not do this in the wild means nothing less than that their arboreal environment makes no demand on them to do it. Correspondingly, they have never evolved the motor control of the throat that is seen in the hominids, though chimpanzees have great natural ability to communicate by gestures. They cannot develop spoken language because they are physically incapable of pronouncing words.

Thus the demands of the savannah environment were probably responsible for the development of intelligence and technological society in man. But it does not follow that this type of environment is a necessary prerequisite to the development of these characteristics. Bernard Campbell thought it was crucial that an animal well adapted to life in the complex forest environment proved to be pre-adapted to an ecological niche on the savannah and was successfully able to invade it. This opened many new possibilities. Moreover, any animal is likely to expand its range into new environments given the necessary amount of preadaptation. Thus what appears most important in stimulating evolution is the presence on a planet of a large diversity of environments readily accessible to the inhabitants of each. The larger number of possibilities inherent in this situation will result in the greatest diversity of species, and will speed the development of traits having survival value, such as intelligence.

This requirement allows us to characterize planets likely to produce intelligent, technological civilizations. In the first place they must have heterogenous and time-variable environments. Bernard Campbell pointed out that in a very stable, homogenous environment no evolution at all occurs, even over indefinitely long time scales. On Earth the deep-sea echinoderms are evidence of this. John Eisenberg pointed out that on isolated land masses intelligence appears to have developed very slowly. For example, in Australia, Madagascar, and pre-Pliocene South America, there was development of many mammalian species, independent of that which took place on the contiguous land masses. On none of the isolated land masses did there develop mammals with large cranial capacity. We see this today in the mammals of Australia and Madagascar. This was also true in South America, although the development of these species was interrupted in the Pliocene, when the northern and southern continents were joined. Only mammals on large contiguous land masses developed both large bodies and large cranial capacities. It is unlikely that planets with limited contiguous land area will evolve intelligent terrestrial life very rapidly, while an aqueous environment is not conducive to the evolution of manipulative ability and hence technology. However intelligent dolphins may ultimately prove to be, they will never build radio telescopes or spacecraft.

Almost certainly once a species with the requisite intelligence, manipulative ability, and complex social organization has evolved, technological civilization will develop. Modern man is little different biologically from Cro-Magnon man. To go from a stone age culture to our present level of technological development required no biological evolution. All that was needed was the development of ideas, and their testing by trial and error. Philip Morrison pointed out that Turing had shown that once yes-no choices can be recognized in large numbers, one can program any mathematical computation, given enough time. Thus, once a system capable of conceptualizing sophisticated internal models of external phenomena has evolved, it is only a matter of time before all possible ideas inherent in the available sensory perceptions are conceived. It is incorrect to focus on "critical" historical events in cultural evolution, just as it is incorrect to focus on single steps in biological evolution. For example, the importance of what might have happened had the Greeks lost at Marathon has probably been greatly overemphasized. Some developments that subsequently occurred might, to be sure, have been prevented, but others of similar type that did not happen could also have been stimulated.

The next stage in the evolution of hominids could be stimulated by our entry into space, including a search for extraterrestrial intelligence. Our past success has been due to our breaking of new ground – to our acceptance of the challenge inherent in exploring new ways of life. This was true even for our distant ancestors who abandoned the relative security of the trees to compete with other carnivores for fleet-footed prey on the open savannah. Hans Lukas Teuber thought the urge to explore and seek new understandings is among our most powerful innate drives. This can only be true because exploration has had high survival value to us.

Thus it might be a mistake if, in our present time of environmental, political, and social crises, we turned our back on that great arena we have not explored – the stars. Daniel Kevles, historian from the California Institute of Technology, said he viewed the matter as a religious man. He noted that everybody's assumption seemed to be that a search was worth conducting if it appeared that life were common. However, he thought it would be worthwhile should we not be convinced of this. A null result would be important. The justification of the experiment is that it is a very special test of whether we are alone; this goes beyond the benefits we may gain or the knowledge we could acquire. Only if we can demonstrate that the probability of intelligent life elsewhere is zero, should we not go ahead with the search; this, of course, is impossible, because we are here. Jack Harlan, historian of agriculture from the University of Illinois, thought we must learn what we can about the Universe; it is our destiny.

Prepared by: Mark A. Stull
SETI Program Office
Ames Research Center

3. DETECTION OF OTHER PLANETARY SYSTEMS

INTRODUCTION

Since we are the only intelligent life we know of, we generally assume, as mentioned in the Introduction, that whatever intelligent species may exist elsewhere also originated on a planet. If the quest for other intelligence is to succeed, the fraction of stars with planetary systems must be reasonably large.

What do we know about this fraction? Extraordinarily little, and the little we know is clouded by controversy. A near neighbor (Barnard's star) was supposed to show small perturbations induced by planetary companions. Refinements of astrometric technique have recently shown these perturbations to be in serious question. No other astrometric perturbation was as large; as a consequence we do not know whether any planets, other than our own neighbors, exist in the Galaxy. Is our solar system unique?

Is this state of knowledge concerning the frequency of occurrence of planetary systems likely to change in the near future, or must we pursue a search for extraterrestrial intelligence in the absence of basic data? The answer depends on a willingness to invest time, thought, and money in an effort to overcome our ignorance. A first step, an investment of time and thought, is under way. A group of scientists (see Section III-15) under the leadership of Jesse Greenstein, has attempted to define how observations might shed some light on the frequency of low-mass companions to stars. A report to NASA will be based on the two Workshops on Extrasolar Planetary Detection held at Santa Cruz (March 24–25, 1976) and at NASA Ames (May 20–21, 1976), and on a meeting of the astrometric community (U.S. Naval Observatory, May 10–11, 1976). This report will outline technical problems and a coherent program that might produce answers. The time scale for detection is not short, since characteristic indirect methods involve years (planetary revolutions) and direct methods may require an orbiting space telescope. Classical means for detecting low-mass companions to stars involve accurate positional astrometry, useful for nearby stars, or accurate determination of radial velocity changes. Both techniques can be improved with modern technology. Direct detection, at optical or infrared wavelengths, is also subject to orders-of-magnitude technological improvement. A listing of science-related activities associated with detection of other planetary systems is presented in Section III-6.

INDIRECT METHODS

If a star has a companion, the star will revolve about the center of mass of the system at the angular velocity of its dark companion. The radial component of velocity is observed spectrographically (independent of distance from the observer), the tangential component is observed with respect to an inertial frame, defined by other stars, as a sinusoidal term superposed on the tangential motion of the star (decreasing with distance from the observer).

Radial Velocity Techniques

What level of accuracy is required to detect the radial velocity effect arising from the orbital motion of a star around the barycenter of a star-planet system? Jupiter's motion around the barycenter of the solar system causes a reflex movement of the Sun of approximately 12 m/sec (with a period of 12 years). The effect of the Earth on the Sun amounts to about 0.09 m/sec. Thus, to detect Jupiter-like planets around other similar stars by radial velocity determinations we need accuracy on the order of 10 m/sec, while 1 m/sec would be desirable. For Earth-like masses, 0.01 m/sec would be none too high an accuracy if it were possible to achieve.

How far are we from such accuracy? Typical radial velocity measurements are accurate to about 1000 m/sec, well above the level required. However, an autocorrelation system can already attain a few hundred meters per second on faint stars. One such system is that employed by Griffin (Cambridge) and Gunn (Palomar). This system is photoelectric and was designed to work on faint stars rather than give high accuracy. It regularly gives standard deviations of 250 m/sec on stars as faint as tenth magnitude and in integration times of 5 min. Griffin has expressed the view that this technique can eventually produce accuracies of about 10 m/sec. More complicated wavelength calibrations, impressed on the spectrogram, are planned at Arizona and are yet to be tested.

What are the foreseeable limits in terms of accuracy for radial velocity determinations? The general feeling of the Workshop on planetary detection was that a level of accuracy in the neighborhood of 10 m/sec can be obtained if great care is taken. Progress to higher levels of accuracy might be achievable by means of conventional photographic spectroscopy, if certain precautions are met. Such elaborate spectrographs are probably reliable over periods of years, if maintained in a fixed position on a dedicated telescope with an aperture of 1–2 m. Construction time scales for such a radial velocity machine and telescope are short (2–3 years) compared to the detection time (several planetary revolutions, or 2 to 20 years). A fundamental problem is the noise in stellar radial velocities; solar granules have motions comparable to 1000 m/sec, but should average out in integrated light. The Workshop felt that this question should be studied over the next few years, with emphasis on the Sun. We know little about the stability of the solar radial velocity, but magnetographs can study the wave motions and non-radial pulsations, and the work of Hill and others (following Dicke) should determine short-period radial pulsations. The gap in knowledge concerns month-to-year variations in solar diameter, the effect of spottedness, etc. The Workshop felt that a program aimed at determining the stability of the solar velocity is valuable and feasible.

Astrometric Techniques

Another, and perhaps better known way to discover dark companions to a given nearby star is that of astrometric observation. In order to set the level of accuracy we require, it should be noted that the displacement of the Sun due to its motion around the Jupiter-Sun barycenter corresponds to about 10^{-3} arcsec as viewed from a distance of 5 parsecs. The displacement of the Sun due to the Earth is considerably smaller, about 10^{-6} arcsec, again as viewed from a distance of

5 parsecs. With good seeing the disk of a star is one tenth arcsec or larger. Current photographic techniques are static, but new measuring instruments have increased the speed and accuracy of bisection of such images. Present technology gives astrometric precision of about 0.003 arcsec in normal points for a year's observation. At the May conference of astrometers, this accuracy and some ideas for improving it were discussed. There are 15 stars (intrinsically less massive than the Sun) for which this precision suffices to detect the presence of a planet similar to Jupiter in mass and distance from its parent star (in time scales of a decade). No system allows detection of perturbations as small as those produced by planets of Earth-mass.

Future technology to improve astrometric accuracy may involve arrays, for example, of charge-coupled devices, with dimensional stability, high quantum efficiency, and linearity. Turbulence in the Earth's atmosphere is the major factor limiting ground-based astrometry, but interferometry can increase accuracy to 50 μarcsec (according to Currie), even with a 1-arcsec seeing disk. The Workshop agreed that spaceborne telescopes could produce accuracies of a few micro-arcsec. Difficult technological problems need to be overcome before this can be attained. The electronic sensors need to be tested on ground-based dimensionally stable, special purpose astrometric telescopes in good seeing.

It seems clear that an infusion of technology is required in both radial velocity measurements and astrometric measurements in order to be able to reach the levels of accuracy required to detect planetary companions around nearby stars. However, no fundamental reasons preclude these technological advances, in a relatively short time and at modest cost, before they are used in space.

One aspect of the indirect sensing methods is that one must observe the system under study for at least one orbital period of the planetary companion in order to have sufficient information to say with any confidence that an object of planetary mass exists. Because such a long time is required for positive identifications (at least several years), more direct means of detecting planets are desirable as a guide to SETI. The radial velocity technique can be applied to many more stars, but lacks information on the inclination of the orbital plane; it therefore tends to be of a statistical nature. Both techniques will have numerous important scientific offshoots, for example, data on the number of stars of low mass and very low luminosity.

DIRECT METHODS

Detecting planets directly is very difficult. One must rely on either thermal (infrared) emission from the planet, reflected starlight, or line emission characteristic of a planetary atmosphere. In spite of the obvious difficulties associated with each of these three kinds of radiation, direct detection may become feasible by one or more techniques.

What are the problems associated with trying to detect a planet by the light it reflects from its parent star? Consider Jupiter and the Sun. The absolute visual magnitude of Jupiter is about +26, while that of the Sun is about +5. An isolated object of +26 is near the limit of detectability with any telescope, beyond a few parsecs. This magnitude difference of 21 corresponds to a factor of about 2.5×10^8 in brightness. At 5 parsecs, the separation between Jupiter and the Sun is 1 arcsec. The practical problem then is to reduce the brightness of the star's diffraction pattern by some nine orders of magnitude within 1 arcsec of the center of the image. This formidable task has been studied by Oliver. Diffraction theory indicates that if one uses Sonine functions to define an apodization mask, one could reduce brightness by a factor of 10^9 at 1 arcsec from the central image on a one to two meter space telescope of extraordinarily high surface accuracy. Proper image processing will probably allow considerable relaxation in the tolerances needed in the apodized optics.

Detecting thermal radiation from a planet should be relatively easier than optical light. In the Rayleigh-Jeans (long wave) part of the spectrum, surface brightness is linear with temperature and the signal is proportional to area. At a wavelength of 10 μ, Jupiter is only three to four orders of magnitude dimmer than the Sun, a much more favorable situation than in visible light. But detectors at 10 μ are noisier and far less sensitive than their optical counterparts. The Workshop estimated that an isolated object, as bright as Jupiter at a temperature of 300 K, could be detected at 1 parsec using a 1.5-m telescope. The diffraction pattern at 10 μ is larger than the pattern at 0.5 μ in the ratio of wavelengths, that is, 20 times larger. Thus, high resolution is required, as well as low noise level detectors. A space system for infrared interferometry appears quite promising and is well worth study.

An indirect way of detecting planetary systems in the process of formation is by means of their far infrared radiation. Protoplanetary material in the form of dust grains has much larger area, for the same mass, than the resulting solid planets after the dust has condensed. Thus, polarization and infrared observations of young stars in the process of formation may reveal the existence of sufficient quantities of solids in orbital motion to suggest at least the future possibility of planetary-sized bodies. Angular resolution is less important than, for example, polarization measurements of thermal (10–30 μ) radiation showing periodic variability of position angle. Another possibility is the detection of the spectroscopic signatures of solid minerals, near young stars, which would be incompatible with formation in or from stars. Protoplanetary disks seen edge-on may be found by their obscuration of the parent star.

Townes announced at the first Workshop that the Berkeley group has discovered nonthermal CO_2 features near 10 μm wavelength of unexpected intensity, with possibly a small amount of maser-type gain, in the atmosphere of Venus. If nonthermal emission mechanisms can lead to large maser-type gain, the signal-to-noise problem is greatly eased, because the thermal input to the planet is concentrated into a relatively narrow wavelength band. If this emission is not found in the star, and is not a common feature of interstellar gas clouds, then direct detection of the planet may be possible. Are high-gain molecular amplifiers common in planetary atmospheres? The Workshop felt that a list of potential atmospheric masers or lasers should be studied, and an attempt made to see from space whether the Earth (or Jupiter) is masing in some molecular lines. The airglow planetary emission is far too weak to be detected over interstellar distances. The advantage of a planetary maser or laser is that the planet has orbital velocities of several thousands of meters per second, and its periodic variation is therefore easily detected if the maser line has sufficient energy pumped into it.

DUPLICITY OF STARS

Many stars near the Sun are binaries, visually resolved. Others are close enough pairs to be detected by velocity variation. Contraction of interstellar gas clouds down to the dimensions of a star leads to excessive rotation, if the gas has preserved angular momentum. Direct observations show some stars rotating at speeds near breakup. The star may fission into nearly equal stellar masses, the momentum going into orbital motion, or it may leave behind it (as it contracts) a protoplanetary disk, that is, a possible planetary system. Stability of planetary orbits was not considered in detail, but the Astrometric conference thought it clear that if a planet must revolve about a binary star at distances large compared to the separation, it may be too cold. An orbit around one star, small compared to the two stars' separation, may be too hot. A recent radial velocity study of bright F and G stars by Abt and Levy suggested that nearly all stars are members of binary systems. This statement required considerable extrapolation from the actual numbers of stars observed to have variable velocity. Many systems, with periods less than 100 years, clearly were of the fission type, while wide pairs were consistent with independent formation. A rediscussion of the data by Branch suggested that as many as five out of six stars were members of multiple systems. This frequency is also suggested by the scatter in color-luminosity diagrams of clusters. The question of how many single stars (with or without planetary companions) remain is important both for astrophysics and for SETI. Branch's corrections to the findings of Abt and Levy increased this number to 0.15, compared to <0.10 as found by Abt and Levy. In both estimates there is a large gap over which extrapolation is necessary, from the smallest detectable mass in a binary ($\approx 0.10\ M_{\odot}$) to that of Jupiter ($\approx 0.001\ M_{\odot}$).

The most common stars in space are the low mass M dwarfs, which are astrometrically most easily studied. They are subject to flares (bursts of optical, ultraviolet, and probably x-rays and cosmic-rays). Little is known about the incidence and effect of flares, especially among the old M stars. Because they have low mass they are most subject to gravitational perturbations, astrometric or in radial velocity. Much conventional study of M dwarfs can be encouraged with expected large rewards.

CONCLUSIONS

With respect to astrometric techniques for detecting planets, a thorough study of the effect of atmospheric seeing on positional determination should be undertaken, and an examination should be made of possible advantages to be gained by way of electronic detection, as compared to use of photographic plates. The 1976 Ames Summer Study on astrometric technique provided important input on these points. It was also felt by the Workshop that astrometric systems in space would have accuracies at least an order of magnitude better than ground-based systems, but that many technical problems had to be overcome before space-borne systems could become a reality. With regard to radial velocity techniques for detecting planets, it was concluded that a determination of the stability of the radial velocity of integrated sunlight would be very valuable, and that the ideal radial velocity instrument needs to be defined. An attempt should be made to obtain an independent determination of the frequency of binary occurrence, and to examine consequences of a binary system on the stability of planetary orbits. Preliminary bench testing of simple apodizing systems could tell us whether the difficult problem of high mirror surface accuracy can be overcome, for direct means of detecting planets. An effort should be made to determine which planetary molecules might possibly give rise to planetary masers, or other forms of non-thermal emission.

The prospects of increasing our confidence concerning the frequency and distribution of other planetary systems are good, if we are willing to invest the effort. As a consequence of the Workshops, several novel approaches to the problem have come to light, as have potential improvements to classical means of detecting planets.

Prepared by: Jesse L. Greenstein
Professor of Astrophysics
California Institute of Technology

David C. Black
SETI Program Office
Ames Research Center

A photograph of the Orion nebula taken with the Thaw telescope of Allegheny Observatory. The Orion nebula is a region of active star formation at present and contains not only visible stars, but a number of intense radio and infrared sources. The straight line emanating from the star below the nebulosity indicates how that star might move across the field of view if it had no planets, while the sinusoidal line indicates how the star would appear to move if it had a planetary companion which could be detected by precise astrometric observations.

4. THE RATIONALE FOR A PREFERRED FREQUENCY BAND: THE WATER HOLE

Seventeen years ago Cocconi and Morrison (ref. 1) suggested that we search at frequencies near the hydrogen line for signals emitted by advanced extraterrestrial civilizations attempting to establish contact with us. At the time, the hydrogen line was believed to be unique but, since then, dozens of other microwave emission lines from a wide variety of interstellar molecules have been discovered. In 1971 the Cyclops study (ref. 2), for reasons that are believed to be rather fundamental, identified the band between 1400 and 1727 MHz bounded at the low end by the hydrogen line (1420 MHz) and at the high end by the hydroxyl lines (1612 to 1720 MHz) as a prime region of the spectrum to be searched for interstellar signals. Because of these limiting markers the Cyclops team dubbed this region the "water hole" and suggested that different galactic species might meet there just as different terrestrial species have always met at more mundane water holes.

At present there is no serious interference in the water hole but navigational satellites and other systems are being planned that would fill the band with interfering signals such as continuous pseudo-random wide band noise. If these systems become operational as allocated, a substantial fraction if not all of the water hole may be rendered unusable for the search. The proposed services can be shifted to other frequencies without appreciable loss of effectiveness but, if the rationale for the water hole is correct, the search for intelligent extraterrestrial life cannot. It would be a bitter irony if the desire to know exactly where we were at all times on Earth were to prevent us from ever knowing where we are with respect to other life in the Galaxy. It is therefore timely to reexamine the case for the water hole in order that we do not, out of ignorance or carelessness, forever blind ourselves to the signals from advanced societies (see Sections III-8 and III-9).

The basic premise that leads us to the water hole is that any advanced society wishing to establish contact will choose the least expensive means that will nevertheless ensure success. As we shall see, one of the dominating factors is the energy that must be expended by the society to announce its existence over interstellar distances, not just to us but to all likely planetary systems. It is this consideration that leads to the radiation of electromagnetic waves rather than probes or spaceships and to the spectral region of the water hole.

THE CASE FOR ELECTROMAGNETIC WAVES

In all probability we will have to examine thousands if not hundreds of thousands of targets before we succeed in detecting intelligent life. The nearest civilization is probably several tens and maybe hundreds of light years from the Sun. With such a profusion of targets, all at such enormous distances, it appears that search by actual manned space travel or by sending out swarms of probes is out of the question (see Section III-1). This approach is far too consumptive of time and energy.

Just to send *one* spaceship to a nearby star and return it in twice the round trip light time, using not fusion or fission power with low yields but matter-antimatter annihilation, would require $\sim 10^{24}$ J. This is enough energy to supply the present total U.S. electrical power for 50,000 years or to keep a 1000 MW omnidirectional beacon shining for 10 million years. Such a beacon could be received by civilizations around any of the million or more F, G, and K stars within 1000 light years, and would probably not need to be radiated by the searching society for more than 1000 years. If so, the beacon is 10^{10} times as effective per joule.

If we seek to reduce the energy requirements of space travel by, say, 10^4, we incur round trip travel times of 400 years per light year or 1600 years for the nearest star. Clearly we are not going to find other intelligent life by hurling tons of matter through space but by receiving – and possibly some day sending – some form of radiation.

Regardless of what form of radiation is used, in order to detect that a signal exists and that it is of artificial origin:

1. The number of particles received must significantly exceed the natural background count
2. The signal must exhibit some property not found in natural radiations

In addition the radiation should

3. Require the least radiated power
4. Not be absorbed by the interstellar medium or planetary atmospheres
5. Not be deflected by galactic fields
6. Be readily collected over a large area
7. Permit efficient generation and detection
8. Travel at high speed
9. Normally be radiated by technological civilizations

Requirements 1, 3, and 8 taken together virtually exclude from consideration all particles except those having zero rest mass. The kinetic energy of an electron travelling at half the speed of light is 10^8 times the total energy of a 150 GHz photon. All other factors being equal the electron communication system would require 100 million times as much power. Baryons are worse. In addition, all charged particles fail requirements 4 and 5. Of the zero rest mass particles, gravitons and neutrinos fail requirements 6, 7, and 9. Of all known particles, only low energy photons meet all the listed criteria. It is almost certain that interstellar communication, if it exists, is accomplished by electromagnetic waves.

We believe that requirement 2 is satisfied by any spatially and temporally coherent electromagnetic wave. Modulation of such a wave in order to transmit information need not destroy the coherence to such an extent that it would be mistaken for a natural signal. Further, the modulation will very likely contain regularities and repetitions of complex patterns that are inexplicable from natural sources.

THE OPTIMUM SPECTRAL REGION

Electromagnetic radiation covers a practically unlimited frequency range. Unless we can find some reason to prefer a relatively narrow portion of the spectrum, one of the dimensions of the search space remains unbounded. Here again we can appeal to energy minimization.

Let us examine critically what is implied by requirement 1 above (that the number of particles received exceed significantly the natural background count). First let us assume that we are operating in the "quantum" region of the spectrum (i.e., $h\nu/kT > 1$), where direct photon detection is easy, and that there is no background radiation. In principle, we require at least one received photon in the observing time τ in order to detect a signal at all. This requires that the received power

$$P \geqslant \frac{h\nu}{\tau} \tag{1}$$

so, for a constant collecting area, the equivalent isotropic radiated power (EIRP) must be proportional to frequency. In practice the reception of a single photon would hardly convince us that we had detected extraterrestrial intelligence. In order to determine the coherence and other properties of the signal we would need to receive some number, n, of photons. But n depends only upon the sophistication of our data processing and upon our assigned a priori improbability and does not depend upon the operating frequency. Thus the proportionality implied by equation (1) still holds, as long as there is no background.

If the radiated signal is coherent and of constant amplitude, the photon arrivals will be Poisson distributed. If the expected number of arrivals in the observing time τ is n, the mean square fluctuation will also be n. The detected signal-to-noise *power* ratio (DSNR) is the ratio of the square of the mean to the mean square fluctuations; that is,

$$\text{DSNR} = \frac{\bar{n}^2}{\bar{n}} = \bar{n} = \frac{W}{h\nu} = \frac{P\tau}{h\nu}\ , \tag{2}$$

where W is the total energy received during the observation and P is the instantaneous power. In communication systems τ is the Nyquist interval and is the reciprocal of the RF bandwidth B. Then equation (2) becomes

$$\text{DSNR} = \frac{P}{h\nu B} \tag{3}$$

and the photon shot noise behaves as if it had a spectral power density $h\nu$.

The above considerations suggest that the lower the operating frequency the less will be the required EIRP and this is true until we reach the "thermal" region (i.e., $h\nu/kT < 1$). In this region direct photon detection becomes difficult but the "radio" techniques of linear amplification, mixing, and spectral analysis are applicable; in fact, these have by now been extended into the optical region. Suppose we Fourier transform a block of the received IF signal of duration τ. In effect, a constant amplitude monochromatic signal is then detected as a rectangular pulse of duration τ to which noise has been added. The effective selectivity curve of the Fourier transform will be the transform of the time window or

$$F(\omega) = \frac{\sin(\omega - \omega_o)\tau/2}{(\omega - \omega_o)\tau/2} \tag{4}$$

and this is the matched filter through which a *CW* pulse of frequency ω_o and duration τ should be passed to give the greatest ratio of peak output signal power to mean square fluctuation in the peak. For this case, or for any equivalent optimum detection process, the detected signal-to-noise ratio is

$$\text{DSNR} = \frac{W}{\Psi} = \frac{P}{\Psi B} \tag{5}$$

where Ψ is the spectral noise power density.

Three contributions to Ψ are unavoidable and are essentially the same anywhere in the Galaxy; these are shown in figure 1 (divided by k to give their equivalent noise temperature). The first is the synchrotron radiation of the Galaxy itself, given by

$$\Psi_g \sim kT_o \left(\frac{10^9}{\nu}\right)^{2.5} \tag{6}$$

where the coefficient T_o varies from about 1 to 2.5 depending upon galactic latitude. This noise dominates below 1 GHz but, above this frequency, rapidly becomes less than the relict cosmic background radiation:

$$\Psi_b = \frac{h\nu}{e^{h\nu/kT} - 1} \tag{7}$$

where $T \sim 2.76$ K. Finally, for $\nu > kT/h \sim 60$ GHz the spontaneous emission noise

$$\Psi_o = h\nu \tag{8}$$

becomes dominant.

The sum of these noise contributions defines a broad quiet region extending from about 1 to 60 GHz known as the *free-space microwave window*. On Earth (or on any Earth-like planet) the absorption lines of the water vapor and oxygen in the atmosphere re-radiate noise with broad

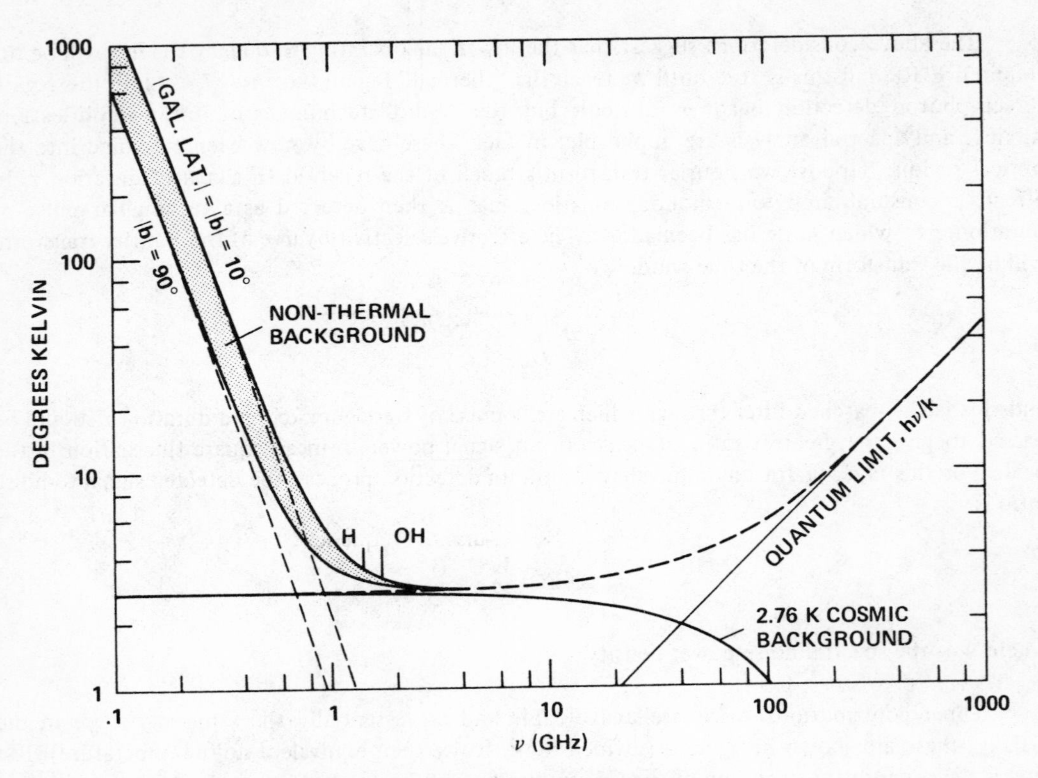

Figure 1.– Free space microwave window.

peaks at 22 and 60 GHz that degrade the window above about 10 GHz, as shown in figure 2. Thus the *terrestrial microwave window* extends from ~1 to ~10 GHz, and is clearly a preferred region of the spectrum for interstellar search using ground-based low-noise receivers.

The suggestion is sometimes made that one might avoid Ψ_g and Ψ_b by going to very high frequencies and then eliminate Ψ_o by employing direct photon detection. At frequencies for which Ψ_o dominates, equation (5) becomes

$$\text{DSNR} = \frac{P}{h\nu B} \tag{9}$$

and, since B can, in principle, be made as narrow as $1/\tau$, we find

$$\text{DSNR} = \frac{P\tau}{h\nu} \tag{10}$$

exactly as given by equation (2). Thus, ignoring technological limitations, linear amplification is just as good as direct photon detection. Spontaneous emission noise does not make linear amplifiers noisier than photon detectors, it merely prevents them from being quieter, which would violate the principle of complementarity (ref. 3).

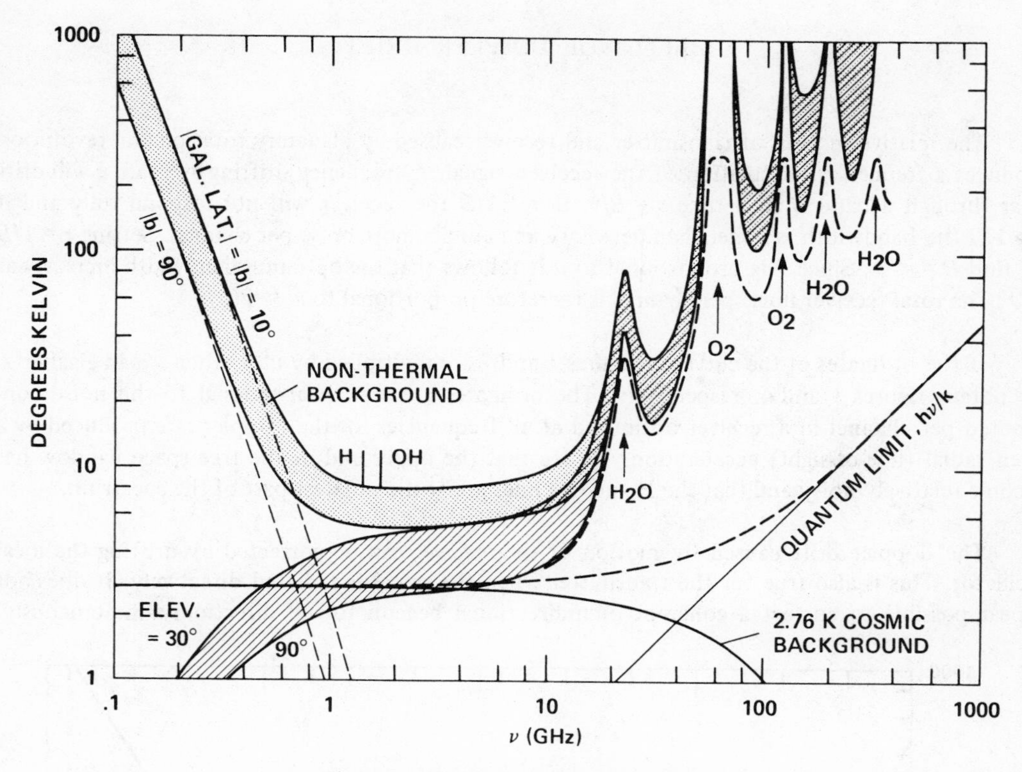

Figure 2.– Terrestrial microwave window.

We may be unable, at optical frequencies, to realize bandwidths as narrow as $1/\tau$. The point is that, even if we could, the linear amplifier, which would then be no noisier than the photon detector, would still be far noisier than a linear amplifier operating in the microwave window.

Not only is the minimum detectable received power least in the microwave window, but also the cost per unit of collecting area is less there than for higher frequencies. This latter consideration also favors the low end of the microwave window over the high end. Additional factors favoring the low end are:

- Greater collecting area for the sharpest beam that can be used effectively
- Greater freedom from H_2O and O_2 absorption, which may be stronger on many planets than on Earth if ground-based systems are used
- Higher beacon powers are probably easier to achieve
- Narrower bandwidths are possible

EFFECT OF DOPPLER DRIFT

The relative motion of transmitter and receiver caused by planetary rotation and revolution produces a frequency modulation of the received signal. A frequency drifting at a rate $\dot{\nu}$ will drift clear through a band B in a time $\tau = B/\dot{\nu}$. If $\tau < 1/B$ the receiver will not respond fully and if $\tau > 1/B$ the bandwidth is larger than necessary and admits more noise per channel. Setting $\tau = 1/B$ we find $B = \nu^{1/2}$. Since $\dot{\nu}$ is proportional to ν it follows that the optimum bandwidth increases as $\nu^{1/2}$. The total receiver noise per channel is therefore proportional to $\nu^{1/2}\Psi(\nu)$.

If the ordinates of the curves of figures 1 and 2 are multiplied by $\nu^{1/2}$ when ν is in gigahertz, we obtain figures 3 and 4, respectively. The ordinates are now proportional to the noise contributed per channel in a receiver optimized at all frequencies for the Doppler rate produced by a given radial (line-of-sight) acceleration. We see that the upper end of the free space window has become relatively noisy and that the H and OH lines are at the quietest part of the spectrum.

The doppler drift caused by motion of the receiver can be corrected by drifting the local oscillator. This is also true for the transmitter when the signal is radiated directively. It does not appear possible to correct a coherent omnidirectional beacon for all directions simultaneously.

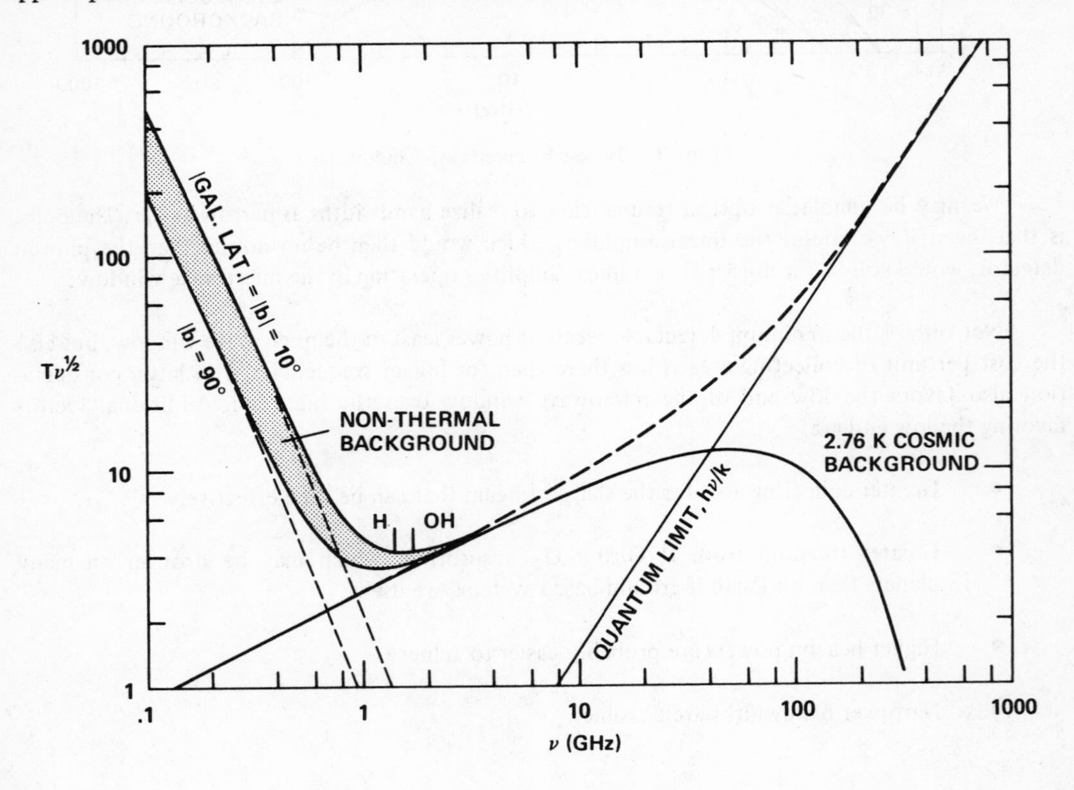

Figure 3.– Free space temperature bandwidth index.

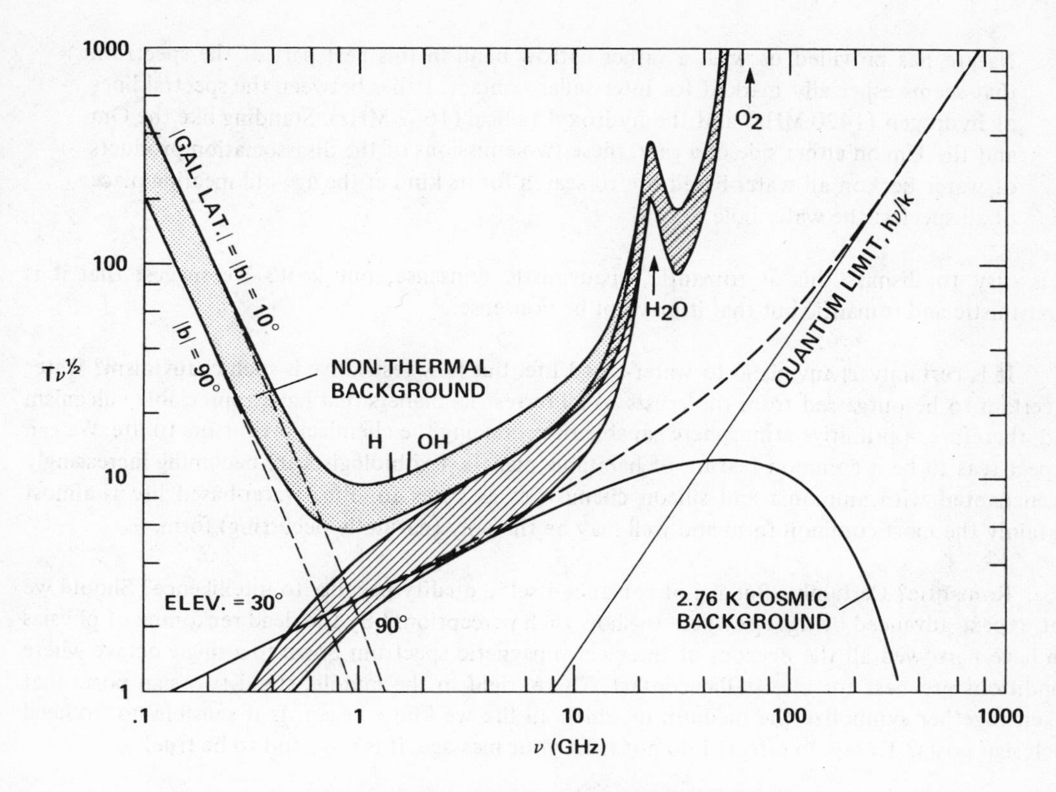

Figure 4.– Terrestrial temperature bandwidth index.

One can of course put such beacons at the planet's poles or in long period orbits in space. Whatever is done to minimize Doppler and oscillator drifts, the frequency proportionality remains. The absolute achievable bandwidth at any frequency may be reduced by improved technology but the frequency of the minima in figures 3 and 4 is unaffected.

THE WATER HOLE

Even though minimum receiver noise per channel can be achieved at about 1.5 GHz, as shown in figure 2, the increase is not very rapid on either side. We are left with at least a 2-GHz-wide region of the spectrum with very little technical reason to prefer one part over another. Clearly, this band is too wide to reserve for interstellar search purposes, being needed for other services (see Sections III-8 and III-9). It was here that the Cyclops team, seeking to economize on search time and spectrum occupancy, observed that the hydrogen and hydroxyl lines are right at the optimum spectral region and, between them, define a rather conspicuous band. As stated in conclusion 6 of the Cyclops report:

> Nature has provided us with a rather narrow band in this best part of the spectrum that seems especially marked for interstellar contact. It lies between the spectral lines of hydrogen (1420 MHz) and the hydroxyl radical (1662 MHz). Standing like the Om and the Um on either side of a gate, these two emissions of the disassociation products of water beckon all water-based life to search for its kind at the age-old meeting place of all species: the water hole.

It is easy to dismiss this as romantic, chauvinistic nonsense, but is it? We suggest that it is chauvinistic and romantic but that it may not be nonsense.

It is certainly chauvinistic to water-based life, but how restrictive is such chauvinism? Water is certain to be outgassed from the crusts of all terrestrial planets that have appreciable vulcanism and, therefore, a primitive atmosphere capable of producing the chemical precursors to life. We can expect seas to be a common feature of habitable planets. Exobiologists are becoming increasingly disenchanted with ammonia and silicon chemistries as bases for life. Water-based life is almost certainly the most common form and well may be the only (naturally occurring) form.

Romantic? Certainly. But is not romance itself a quality peculiar to intelligence? Should we not expect advanced beings elsewhere to show such perceptions? By the dead reckoning of physics we have narrowed all the decades of the electromagnetic spectrum down to a single octave where conditions are best for interstellar contact. There, right in the middle, stand two sign posts that taken together symbolize the medium in which all life we know began. Is it sensible not to heed such sign posts? To say, in effect: I do not trust your message, it is too good to be true!

In the absence of any more cogent reason to prefer another frequency band, we suggest that the water hole be considered the primary preferred frequency band for interstellar search. This does not mean that other frequencies should be ignored. Harmonics of the hydrogen line deserve some attention. In space, the waterline itself, at 22 GHz, may have merit. It does mean, however, that the water hole deserves the greatest attention for protection against interference.

It is always possible to dismiss the above argument on the grounds that we do not know everything yet and that there may be some more compelling reason to choose another frequency band or even some as-yet-to-us unknown method of communication. These assertions are undeniable and also unacceptable since they logically lead to never doing anything. If we are to make progress we must proceed on the basis of what we know, and not forever wait for something now unknown to be discovered.

REFERENCES

1. Nature 183, 844, 1959.

2. Oliver, B. M., and Billingham, J.: Project Cyclops, A Design Study of a System for Detecting Extraterrestrial Intelligent Life. NASA CR 114445, 1972.

3. Heffner, H.: The Fundamental Noise Limit of Linear Amplifiers. Proc. IRE, vol. 50, no. 7, July 1962, pp. 1604–1608.

Prepared by: Bernard M. Oliver
Vice-President of Research and Development
Hewlett Packard Corporation

5. SEARCH STRATEGIES

INTRODUCTION

The objective of SETI strategies is the detection of the existence of other intelligent species in the Universe by examining the spectrum of electromagnetic radiations in the vicinity of the Earth. Communication, one-way or two-way, does not directly concern us now, only detection.

Another characteristic of this objective is that out of all the intelligent species that may exist, we are seeking another member of the subset to which we belong. The essential property of our subset is that its members radiate electromagnetic energy into interstellar space in such large amounts and in such a distinctive fashion that at interstellar distances we can recognize it as an artifact against the background of "natural" radiations.

In the past several decades we have developed a trenchant technology appropriate to searching in the spectral range of the free-space microwave window. No comparable technology has yet been generated for shorter wavelengths, so the remarks below are directed in the main toward strategies suited to an initial search in the microwave window and give special attention to the water hole (see Section II-4). However, our suggested strategies do encompass the spectrum above the microwave region, for relevant physical knowledge and suitable technologies are burgeoning (e.g., in the infrared region of the spectrum).

A final distinction – search strategies as discussed here, though closely related, are not equivalent to search programs. The latter are a topic unto themselves since many factors not touched on here enter their design.

GENERAL CONSIDERATIONS

We cannot afford to search for all kinds of radiation at all frequencies from all directions at the lowest detectable flux levels. It is necessary to put some bounds on the volume of our multidimensional search space. At the same time it is important not to narrow the space too much: to put all our eggs into one basket. We submit that a rational approach is to assess all strategies and to attempt to assign relative a priori probabilities of success per unit cost. If only one strategy can be pursued at a time, one chooses the most likely and continues until success is achieved or until the accumulated negative results have depressed the probability/cost ratio below that of some other strategy, in which case the other strategy would then be pursued. If several strategies appear to have comparable probability/cost ratios they may all be pursued in proportion to these ratios.

The case for preferring electromagnetic to any other form of radiation seems compelling (see Sections II-4 and III-1). The case for preferring the microwave window (approximately $1-10^2$ GHz) seems very strong but not necessarily compelling. The case for preferring the low frequency end of the window to the high seems strong but not so strong that no attention should be paid to higher frequencies. The case for the water hole is very appealing (as a starting place) if one has already decided on the low end of the microwave window.

The case for searching nearby main sequence F, G, and K stars at ever-increasing range seems very natural; the only life we know lives on a planet around a G2 dwarf star. To adopt this strategy takes us only as far into space as necessary to find perhaps our closest neighbor. Communication with a close neighbor would permit more two-way exchanges than with a civilization in the Andromeda nebula, for example.

On the other hand, cultures only slightly older than ours may be able to exploit enormously greater communicative capabilities. As is true for stars, the nearest transmitters may not be the strongest. The strongest signals may come from more advanced cultures at great distances. For these reasons it would be a mistake to pursue only one search strategy, such as that suggested in the Cyclops report. One should in addition examine other options. To cover other possibilities, it seems prudent to conduct a complete search of the sky over as wide a frequency band as practicable (see Section III-3). To be significant, such a search should extend to frequencies and down to flux levels not reached by existing radio astronomy surveys.

Our great uncertainty as to the likely flux levels of extraterrestrial signals argues that all search strategies should assume at the outset the strongest signals not likely to have been detected yet, and that the sensitivity should be increased with time until success is achieved or until the strategy is no longer thought to be sufficiently promising.

PARTICULARS

The fund of relevant physical knowledge and available technologies has a profound effect on estimates of a priori probability of success per unit cost of a proposed strategy. This is particularly true with respect to individual subelements. Furthermore, certain concepts may be unattractive simply because they promise significant returns too far in the future even though the total costs might be relatively low. Human time scales and human patience are important factors in the present context.

The Satellite Problem

It is difficult to survey the entire sky over the entire $1-22$ GHz spectral range from the surface of the Earth. The glare of satellite transmissions over more than half this band makes it difficult to reach attractive sensitivities in the satellite bands (see fig. 1).

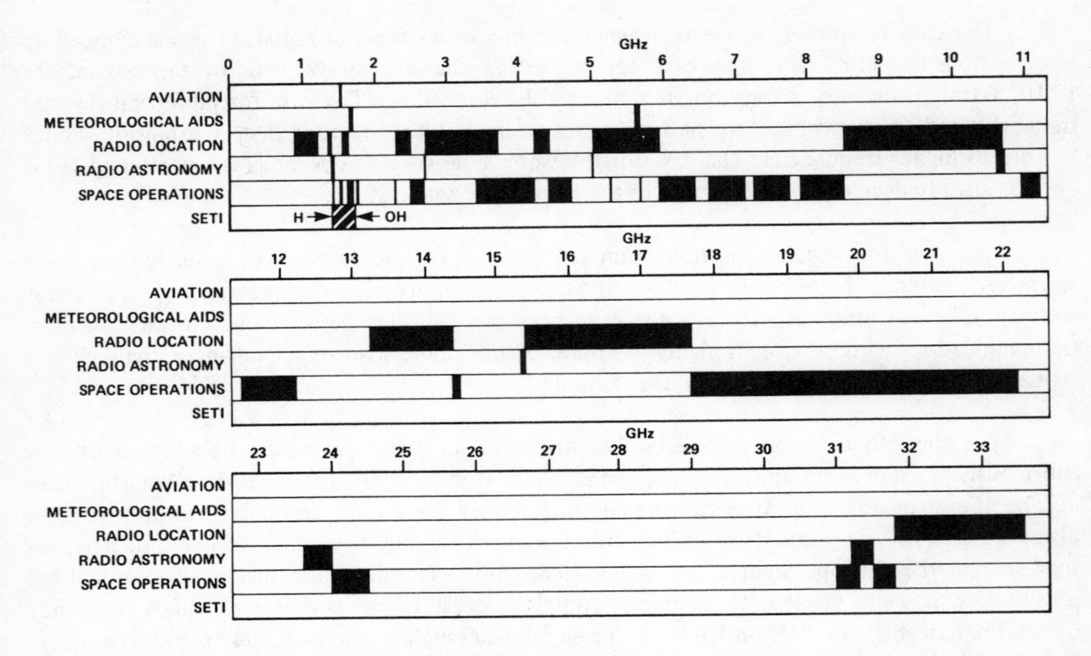

Figure 1.– Frequency bands in the free-space microwave window, allocated to certain transmitting services and the "water hole" allocation recommended for SETI.

Radio Astronomy Surveys

Radio astronomers have carried out few surveys of the sky to what are believed to be flux levels of interest here. It is even possible that one or more of the "point" sources appearing in a survey in one frequency band but not in another survey in a different band, could be an unexpectedly powerful artificial signal.

Radio astronomical surveys of large areas of the sky can be divided into two classes: broadband continuum surveys (1 MHz to 5 GHz bandwidth), and spectral line surveys (10^2 to 10^6 Hz bandwidth). Large bandwidths, absence of adequate spectral resolution, customary on-line instrumental data compaction, and absence of specific interest in the existence of extraterrestrial artifacts, all have reduced enormously the chances of recognizing an artificial extraterrestrial signal. In fact, a signal so strong as to be noticed, generally would have been ascribed to man-made radio frequency interference (RFI).

To illustrate these points, we describe briefly two surveys reaching low flux levels by radio astronomical standards. The Parkes 2700-MHz source survey of the sky visible from Australia employed a bandwidth of 200 MHz and a sensitivity such that a signal flux level of about 10^{-20} W/m^2 would have produced unity signal-to-noise ratio (SNR = 1). No procedures were in effect to recognize a narrow band signal at this or much higher levels, unless the absence of this "radio source" on other surveys at adjacent frequencies caught an astronomer's interest and the source was then reexamined with suitable care.

Westerhout used the NRAO meridian transit 300-ft antenna to survey a region along the Galactic equator 225° in longitude by 4° in latitude. Frequency resolution was 9.5 kHz in a band about 1 MHz wide. Unity SNR corresponded to about 3×10^{-23} W/m^2, but automatic data handling algorithms and RFI subtraction procedures would generally have prevented output registration of a coherent signal orders of magnitude stronger.

The radio astronomy situation (and parallel situations in other areas) can be summarized this way. Procedures in use in high data volume observations tend to discriminate against the discovery of unexpected phenomena. Scientists generally do not explore a range of phenomena merely because it is not forbidden by any known laws. They usually have more immediate objectives in mind. The polarization of starlight was observed half a century after the work of Maxwell and Hertz; and the discovery of pulsars was a fortuitous accident due to the presence and curiosity of an astute graduate student in a favorable observational situation. Analog pen and paper recording of data was in use. For a third and final example, proof of the polarization of the diffuse background radio radiations was delayed at least 5 years by an unwillingness to test for polarization *just* because it was an unexplored degree of freedom in nature.

Signal Classes

Postulating the characteristics of signals we might be able to detect from another species has enlivened many a casual moment. We have only our experience over the past 30 years or so, our electromagnetic technology in which we take some pride, and our projections of how in the near future we might exploit this technology to suit our manifold desires.

Above some tens of megahertz most of the power from all our surface transmissions is dissipated in the endless reaches of outer space. For over a quarter century and with increasing intensity, we have been generating an expanding aura of fairly powerful signals about the Earth, one *we* could detect if we had a sufficiently sensitive radio telescope situated 20 light years from Earth. As has been suggested many times, this may be a transient phase in our development, one that may be over in a time period microscopic on the cosmic scale. Nevertheless, a reversal of the historical trend toward higher transmitter powers is not yet in evidence. It does seem likely that some of our present transmissions will vanish from our scene in favor of more efficient, more directive procedures; cable and low power satellite transmissions may replace high power TV broadcasting, for instance. The future of high power radar is perhaps less clear. Table 1 is a fairly complete tabulation of published current high power radar signals in the 1 GHz to 2.5 GHz frequency range. They generally sweep out the sky continuously in regular patterns both in time and in direction.

One class of signal for which we might search is the intraspecies type of transmission. We may be fortunate in having a relatively close neighbor on whom we can eavesdrop. From our own experience, this would require not only proximity or exceptionally great sensitivity because of the modest power levels we may perhaps expect, but also a complex pattern recognition capability in the time, frequency, and modulation domains. Most of our transmissions have a strong carrier component. If this holds with our nearest neighbor (and it may because it is a simple way to

TABLE 1.– HIGH POWER TERRESTRIAL RADIATIONS IN THE 1–2.5 GHz RANGE[a]

EIRP, W[b,c]	No. of transmitters[d]
$>10^{12}$	1
$10^{11} - 10^{12}$	2
$10^{10} - 10^{11}$	0
$10^{9} - 10^{10}$	15
$10^{8} - 10^{9}$	23
$10^{7} - 10^{8}$	78
$10^{6} - 10^{7}$	462
$10^{5} - 10^{6}$	2680

[a]This table is a summary of published radar installations.

[b]Equivalent Isotropic Radiated Power in watts.

[c]UHF color TV (460–890 MHz provides many carriers in the 10^5–10^6 W range. They are radiated with high stability and the toroidal radiation patterns rotate with the Earth. Television signals may constitute the most readily detectable terrestrial radiations at moderate distances in spite of their lower frequency.

[d]See table 3 in Section III-2 for further details.

maintain coherence), it eases the pattern recognition problem since stable, monochromatic signals are relatively easy to detect. Finally, we are in total ignorance of *their* spectrum utilization and have little idea even how they might utilize transmitters in "space activities." We can only assume that bandwidth and propagation requirements dictate some frequency allocation scheme there as here. We have, so far, no reason to search any frequency band outside the low noise, free-space microwave window; and since signal stability so directly affects our achievable carrier search sensitivity, the water hole seems a likely place to start an eavesdropping search.

Eavesdropping can be classified as a non-cooperative search situation. Alternatively, "they" may be transmitting in a cooperative, *beacon* mode, carefully arranged to assist discovery by others, particularly newcomers like ourselves. Any concern on their part for range would surely highlight the advantages of the low end of the free-space microwave window. Since we have conceived it and it is feasible for us today, it is quite possible (but who knows how likely) that other species are curious as are we, and indulge in their allotted share of an obvious cooperative search strategy.

Beacons could exist in many modes and for many differing applications. All who study extraterrestrial phenomena should be alert to the possibilities, regardless of the portion of spectrum of immediate interest to them.

It is worth noting that beacons can be of any power, isotropic or beamed, and continuous or repetitive. If I were asked how to construct a beacon (to announce our presence) to be built on a crash basis by the year 2000, I would suggest these chief properties: 1 GW continuous radiated power, an isotropic radiation pattern, a frequency in the water hole some megahertz above 1421 MHz, frequency stability to 10^{-14} or better for the circularly polarized carrier, and modulation by polarization reversal in three modes: (1) carrier alone, (2) a bit every few seconds of binary, "acryptic" information to assist first decoding, (3) large information transfer at a bandwidth up to perhaps 10^4 Hz, and the transmitter and all that it requires in Earth's solar orbit on the other side of the Sun in order to provide a low Doppler drift rate and to minimize pollution of the local terrestrial electromagnetic environment. Modulation modes (1) and (2) would be present over 90 percent of the time in order to assist first detection. Such a beacon installation would be at about the limit of our own technology, and it would be detectable by a system equivalent to a modest Cyclops at a distance of 1000 light years or more.

Signal Propagation Paths

The interstellar medium, the interplanetary medium, and the Earth's atmosphere and ionosphere can all affect the coherence of signals passing through them. The physical processes, dispersion, scatter, and multipath transmission, are well understood theoretically, but our observational knowledge leaves much yet to be learned. In all these media there are systematic and turbulent motions of matter and free-electron density and magnetic field intensity which all show variations in direction, distance, and time. In the lower atmosphere, corresponding variations in the water vapor density add their contributions to the total possible coherence loss. Present information suggests that in the water hole and over distances as great as 1000 light years, the coherence of interstellar intelligent signals may suffer an appreciable loss unless the initial bandwidth (B) is limited to about the range

$$10^{-3}\ \text{Hz} < B < 10^{6}\ \text{Hz}$$

These limits are uncertain by perhaps a factor of 10 and, since bandwidth is an important search dimension, point source scintillation and pulsar pulse-shape studies should be encouraged.

Radio Frequency Interference (RFI)

Solving the RFI problem is crucial to SETI. Section III-8 discusses RFI extensively and Section III-9 presents the unanimous Science Workshop resolution on the matter. Here, we mention briefly only certain salient points.

A SETI system, regardless of antenna area, requires protection from man-made radiations down to a level which is 50 to 100 dB more stringent than the usual communication system requirements (see Section III-8). The protective measures required depend on the location of the SETI system and upon the frequency band or bands being searched.

Figure 1 shows that most of the terrestrial microwave window below 22 GHz has been allocated to services such as radar and satellite communications, and therefore is generally difficult for SETI. It is fortuitous that, so far, the 1400 to 1727 MHz "water hole" band is used chiefly by a multiplicity of low and very low power services with which, in the main, SETI is compatible. SETI does not require an exclusive frequency band allocation. Only a moderate degree of worldwide and local cooperation is needed in order to preserve the terrestrial water hole for SETI. We should:

1. Choose a site for SETI over the horizon from centers of high population density, and out from under heavily used commercial aviation lanes.

2. Avoid the use of water hole frequencies by even low power services within some 100 to 200 km of the SETI site.

3. Obtain worldwide agreement to keep the water hole band significantly free of interference at the one or more SETI sites agreed upon.

4. Allow the present, mild use of the band by satellite services to come to a natural, perhaps hastened, termination over the next 5 or 10 years.

Item 3 involves more than just setting limits on EIRP. All transmitters radiate some power into adjacent bands and into harmonic bands. The official national and international standards on spurious emissions are old and well behind the state of the art; and the actual situation is often worse because of the effort required to challenge effectively operations believed to be below even the standard requirements. Some civil and federal communications groups are already trying to improve both the standards and the practice to levels much closer (at least several powers of ten) to the knee in the performance/cost curve of current first-rate technology.

To summarize the Earth-based search site situation, only relatively mild allocation problems are foreseeable with respect to sharing frequency allocations with many kinds of Earth-based transmitters. Searches cannot compete with line of sight satellite down transmissions. Any overall search strategy should contain an element that proposes actively to support other groups trying to bring transmission practice closer to that permitted by the state of the art. If adequate RFI protection for Earth-based search cannot be provided, it will be necessary to develop space search systems shielded from Earth.

Space-based search systems are without the natural and effective shielding properties of the Earth. In descending order of estimated relative cost, we list below four desirable and feasible (before the year 2000) off-Earth sites for long-term search systems – long-term because much useful initial search and research could be carried out on Earth and in low Earth orbit while developing and testing space-based hardware for more sensitive and extensive systems, if they are needed. The four sites are:

1. On the far side of the Moon

2. In Earth orbit around the Sun and 60° from the Earth as viewed from the Sun

3. At lunar distance at L-3, on the opposite side of the Earth from the Moon

4. In synchronous orbit around the Earth and at the longitude of the data processor

Site 1 seems hardly worth consideration until well after the year 2000, for reasons of cost. International efforts are already under way to keep the electromagnetic environment essentially pure as seen from the far-side lunar surface. The Moon itself is an excellent shield against terrestrial radiations.

Site 2 employs distance rather than an RFI shield, to protect itself from Earth's radiations. It is likely to be more costly in the long run chiefly because of servicing costs, reliability requirements, and multiple, long distance relay link demands. Also, some minor but necessary limits on maximum permissible EIRP on Earth would be required in the search frequency bands. The greater distance provides only about 50 dB improvement over Site 3.

Site 3 requires either a shield against Earth and Earth satellite radiations or a nearly sole allocation of the search frequency bands to the search effort. A shield would allow search of the entire free-space window and pose no allocation requirements. However, RFI shields could be costly.

Site 4 can tolerate almost no Earth or satellite transmissions in the search bands, unless an RFI shield is used. Even then, there would likely be restrictions on satellite transmissions in the search bands. A careful engineering assessment would be required of any concrete proposal involving this site.

The RFI shields mentioned above are worth brief description. They are needed because realizable antennas are imperfect in the sense that there is no direction of signal arrival to which they are totally unresponsive. The purpose of the shield then is to attenuate signals by reflection, by absorption, and by diffractive loss, such that the signal energy reaching the antenna does not produce a detectable signal at the antenna output terminals.

Precise definition of shield requirements in any particular instance depends on the nature of the interfering fields (direction and strength) and on the directional characteristics of the antenna. The latter are variable, of course, since the antenna is required to look in many directions relative to the direction of the source of interference. Typically, in situations studied so far, shield attenuations on the order of 50 to 200 dB (i.e., power ratios of 10^5 to 10^{20}) are required. To visualize the realization of such a shield, imagine a 300-m space antenna in Earth orbit. Adjacent to it and between it and the Earth is a large disk consisting of a ring supporting a thin conducting membrane. The disk diameter would be of the order of 450 to 600 m. The conducting membrane may be only several tens of microns thick; even be a fine conducting mesh. The whole is equipped with means for orbital station keeping, engines, sensors, telemetry, etc., so that it rotates about the Earth in step with but unattached to the antenna. A smaller or larger antenna would require a proportionately smaller or larger shield.

Because such shields have not yet been developed, the need for thorough study is obvious. It is also possible that the shield may somehow be incorporated into the structure of the antenna itself.

Near-Earth orbit sites are omitted from the list because (as argued later on) the data processor must be on Earth, and because an antenna moving rapidly relative to the signal processor would seem to present very difficult telemetry problems, considering the SETI need for precise phase and frequency control, and the need for a very great degree of freedom from telemetry and atmospheric noise effects.

Control of miscellaneous interference from neighboring devices such as powerlines, switches, motors, etc., is well understood, but it is not a trivial matter and must not be ignored in the design stages of any SETI system.

In summary, the very real problem of RFI suggests a family of strategies for its alleviation, the nature of a particular strategy depending on a variety of wide-ranging factors such as possible sites, practical antenna characteristics, search frequency ranges desired, cost considerations, etc. There is no easily obtainable "quiet site" in view. On the other hand, the RFI protection problems for some combinations of search system parameters seem to be either minor in nature or at most only a moderate nuisance. This is true because searches are under way now on Earth in some clear bands and can continue for some time before the freedom to search is slowly closed down further by steadily increasing sources of RFI. But the time scale is not tomorrow; rather it is a decade or more in the future. This leaves adequate time for government departments, communication agencies, and so forth to make gradual adjustments toward cooperating with the search effort, thus avoiding exceptional costs and upset plans. Then too, advancing technology hurries obsolescence, so adjusting to the needs of a search system requires mainly thought and willingness, and comparatively small material cost or other inconveniences. It is most fortunate that the next and crucial World Administrative Radio Conference (WARC, see Section III-8) is scheduled for late 1979 and that so far, the water-hole band has not yet been occupied by extensive interference-producing installations such as are present in the bands on either side of the water hole.

Directional Search Modes

There are two distinct directional search modes: (1) the *target search* mode and (2) the *area* or "whole sky" *search* mode. Preference for one or the other, or for a mixture of the two, as major elements in an initial overall strategy for SETI, is a matter of judgment and practicality. Both modes have been and are being utilized. The point of view espoused here favors a modal mix at first, followed by increasing concentration on target search.

Target search assumes the existence of a known set of most likely transmitting sites. For example, in the Cyclops study it was proposed to observe presumed planets around Sun-like F, G, and K dwarf stars, in ascending order of distance from the Sun. There are about 150 of these within 50 light years, but the number increases to about 1.7×10^6 within 1000 light years. Some have argued strenuously that M0-M5 dwarf stars should be included in the target list. This would increase the size of the list by a factor of 2 to perhaps 5. Others have suggested giving first priority to a narrower selection of stars, say, only F5-K5 dwarf stars. Other discrete objects have also been proposed for special study.

The only civilization we know is on a planet orbiting a G2 dwarf star and it is appealing to expect to find other and somewhat similar civilizations on planets revolving around roughly comparable stars. There has been little change in the relevant astrophysical data since the Cyclops study in 1971. As a result, this strategy is widely supported.

Clearly, to minimize effort, target searches should be guided by the latest relevant knowledge in such areas as stellar and planetary formation and evolution, circumstance and origin of life, and cultural evolution. Because of the early stage of our knowledge in some of these fields, an overall search strategy should include a balanced substrategy for increasing relevant background knowledge; such a strategy would likely shorten the time for detection.

At present, astronomical star catalogs can identify perhaps only 0.1 percent of the stars within 1000 light years. A basic catalog listing all stars to 14th or 15th apparent magnitude just has not been constructed. Section III-4 outlines a strategy for generating an adequate Whole Sky Catalog within perhaps a decade and at moderate cost. Such a catalog would be invaluable to general astronomy as well.

In the meantime, the search can proceed by examining the truly local stars which are fairly well cataloged. Our nearest neighbors would seem to deserve rather intense study. Being near, intrinsically weaker radiations are detectable over the entire accessible spectrum. Being few in number, one can afford longer integration times. In any target search scheme, each time sensitivity is noticeably increased, one should reexamine previously observed objects.

The operation of a selected target search is clearly not a simple matter. When the committee or committees, which doubtless will control such matters, sets the following year's list of targets and areas to be searched, a fraction of the observation time should be set aside for trying inspired suggestions, as formal acknowledgment that the establishment is also working in the dark unknown. It is not beyond sensible conception that the first detection will be serendipitous.

Area search specifies nothing about the location of possible transmitting sites in the Universe. It merely characterizes the whole sky as a function of flux level, direction, and frequency.

In Section III-3 an elegant theorem is developed to this effect:

The received (signal) pulse energy is independent of the antenna area and is the energy that would be received by an isotropic antenna in the full sky search time.

This holds as long as all the received data are properly used in the data reduction process. It applies whether the sky is continuously scanned in any essentially nonoverlapping fashion or whether the antenna is used in isolated target search procedures, where a fictitious search time is easily calculated.

From this theorem one can derive an expression for the flux level attainable using an ideal receiver in the real universe and compare its performance with that of a state-of-the-art receiver. This expression is:

$$\phi_o = 4\pi \Psi m / \lambda^2 \, t_s \qquad (\mathrm{W/m^2}) \tag{1}$$

where the flux density, (ϕ_o) varies directly with the system noise power spectral density ($\Psi = kT$), and with the signal-to-noise ratio, (m), required to keep the false alarm rate due to noise peaks at an acceptable level, and inversely with the wavelength, (λ), squared and the full sky search time, (t_s).

Substituting

λ = 0.2 m (1.5 GHz)
m = 25
k = 1.38×10^{-23} J/K
T = 2.7 K or 10 K (ideal or state of the art)
t_s = 0.1 yr = 3.1×10^6 sec

we find

$$\phi_o \text{ ideal} = 3.5 \times 10^{-25} \ \mathrm{W/m^2}$$

$$\phi_o \text{ pract.} = 1.3 \times 10^{-24} \ \mathrm{W/m^2} \tag{2}$$

In perhaps a year's time, given sufficient wideband data processing equipment (see below), it should be possible to conduct a comprehensive search in the radio astronomy bands over the whole sky and with a sensitivity to narrow band signals many orders of magnitude greater than that used in existing radio astronomy observations. There would then be no need for retrospective studies of existing observations and surveys on the chance that radio astronomers have already observed an artificial signal from some fixed direction in space. Again, if we can search a 300 MHz frequency band at one time, it would take only a few years to search the whole sky over the entire microwave window. This would improve the state of our knowledge by many, many orders of magnitude.

Such whole sky area searches are quick and easy to perform with modest (~25 m) antennas, *given the data processing equipment.* They might well detect an ETI signal. Furthermore, an attractive dividend of a search throughout the microwave window would be the characterization of the whole sky at these frequencies to a systematic, known set of flux levels, spatial and frequency resolutions. The resulting astronomical data would be valuable, and it does not seem unduly optimistic to expect new discoveries in the spectral line domain, independent of precise frequency prediction.

Some Technological Aspects

The flux received at the Earth from a transmitter r light years away, per watt of equivalent isotropic radiated power (EIRP) is

$$S = 8.89 \times 10^{-34}\,(\mathrm{EIRP})\,r^{-2} \qquad (\mathrm{W/m^2}) \tag{3}$$

A good receiver has a power sensitivity of

$$P_r = 1.38 \times 10^{-23}\,T_s B_r \qquad (\mathrm{W}) \tag{4}$$

at $S/N = 1$, when the system equivalent temperature lies in the range $2.7\ \mathrm{K} < T_s < 10\ \mathrm{K}$, and the resolution bandwidth (B_r) is equal to or greater than the received signal bandwidth as observed over a unit time $\tau \cong (B_r)^{-1}$.

For purposes of this discussion, assume $T_s = 10$ K and that we are searching for moderately stable carrier signals, so $B_r = 0.1$ Hz. Then,

$$P_r = 1.38 \times 10^{-23} \qquad (\mathrm{W}) \tag{5}$$

The disparity between equations (3) and (5) can only be overcome by some combination of real transmitted power (P_t) and transmission directivity (g_t) ($\mathrm{EIRP} \equiv P_t g_t$) at the source, and effective antenna collecting area (A_e) at the receiving end.

These relationships are illustrated in figure 2 where range in light years (ly) is plotted against effective collecting area expressed in the number of 100-m radio telescopes required. Expressed another way, the range is approximately

$$r = 20[n_{100}(P_t g_t/10^9)]^{1/2}\ \mathrm{ly} \tag{6}$$

where n_{100} is the number of 100-m dishes required if they are 80 percent efficient ($\eta = 0.8$). The horizontal bands indicate the ranges likely needed for conditional detection probabilities in the range $0.63 < p_c < 0.95$ and under four assumptions about the density (N) of transmitting civilizations in the Galaxy.

Antenna area and resolution bandwidth are interchangeable. The collecting area required dominates system costs to such a degree if high flux level signals are absent, that detection

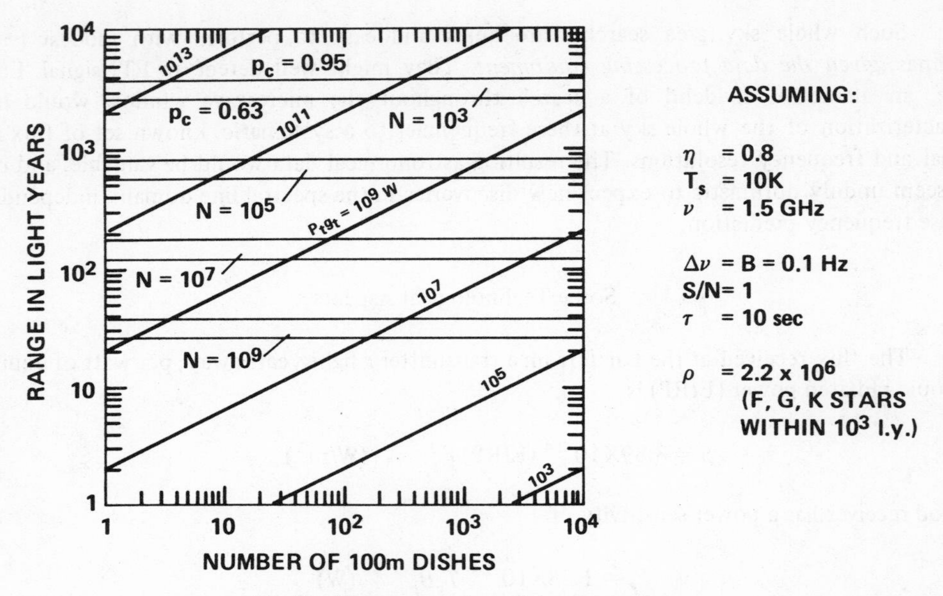

Fig. 2.– Major parameters of signal detection.

becomes, for all practical purposes, a simple matter of carrier recognition. This is so because (so far) we do not have an adequate adaptive matched filter technology for handling more complex coherent signals in a regime where the SNR is close to 1.

Another conclusion is that because optimum carrier detection technology has not yet been brought to bear on the problem, the first instrumental effort should be in this direction. In fact, an improvement by a factor of 10^3 to 10^5 over past and present practice can be expected merely by equipping present radio telescopes with better electronic systems.

Finally, the data processing equipment – Fourier transform spectrum processor and pattern recognition system – should be Earth-based even if the ultimate optimum strategy calls for space-based signal collectors. It is this equipment which is most subject to reorganization as optimum strategies evolve. In addition, we note that unlike maser technology, digital data processing technology is still in an early stage of development. There is, as yet, no quasi-ultimate horizon in view.

SUMMARY OF ATTRACTIVE STRATEGIES

A review of the requirements for search strategies seems desirable and we summarize them here. A rational strategy should:

1. Concentrate on the most likely alternatives and assign proportionally smaller effort to less likely alternatives;

2. Start with the smallest system for which a significant a priori chance of success exists and expand with time until success is achieved or until further effort is felt to be unwarranted;

3. Have substantial objectives which can be achieved within a fraction of the lifetime of the generation that begins the search;

4. Have the least cost for a given probability of success;

5. Produce valuable scientific fallout even in the absence of success (see Sections II-6, III-5, and III-6).

Three strategies are required for the guidance of three corresponding, parallel, and inter-related areas:

1. Exploration of the microwave window, both by target search and by whole sky survey.

2. Development of knowledge in relevant scientific areas.

3. Exploration of the remainder of the electromagnetic spectrum.

In each of these areas, a flexibility and multiplicity of approaches should be positively encouraged. Activity in each area should build up gradually from the present status, starting in each case with a survey of the field and emphasis on the rapid development of obviously key items, followed by an efficient buildup to some generally agreed upon steady state level.

Going one step further into details, the following substrategies are defined.

Exploration of the Microwave Window: Initial Phase

1. Place emphasis on the water hole frequency band at least in the beginning.

2. Concentrate first on carrier search technology.

3. Provide suitable low noise electronic systems for Earth-based operations and develop the equivalent for space systems.

4. Develop $10^6 - 10^{10}$ bin Fourier transform spectrum processors, simple visual, and simple automatic pattern recognition systems.

5. Using target and area search procedures, gain observational experience in characterizing the sky, using items 3 and 4 with existing antennas.

6. Develop and initiate an archival system (see Section III-13).

7. Carry out design studies for ground and space-based experimental, dedicated, small antenna systems, and develop a site-choice strategy.

Relevant Scientific Studies: Initial Phase

1. Plan stellar census construction (see Section III-4).

2. Plan advanced astrometric planetary detection schemes (see Section II-3).

3. Investigate alternates to item 2, particularly direct detection possibilities (see Section II-3).

4. Increase research effort in theoretical and observational investigations of star and planet formation and evolution; in origin of life studies; in pattern recognition; in procedures for recognizing coherent signal statistics under minimum S/N ratio conditions; etc. (see Section II-6).

General EM Spectrum Exploration: Initial Phase

1. Survey the observational needs and strategies required to progressively characterize the entire external electromagnetic spectrum of objects in the solar system and beyond, because signals of intelligent origin could be found anywhere in the spectrum and could be confused with natural background.

2. Begin the development of the most obviously desirable instrumentation for item 1.

3. In view of the number of instances of failure to perceive important new discoveries in data taken for other purposes, and because many discoveries have been serendipitous, measures should be taken to encourage investigators to remain alert to the possibility of ETI artifacts in their data. "Chance favors the prepared mind." – L. Pasteur.

Finally, it is important to note that search strategies should always be evolutionary and quick to respond to new experience, new knowledge, new technology, and to new inspiration. At all stages they should be in full view of humankind, and be a reflection of the spirit and intellect of the entire human species.

Prepared by: Charles L. Seeger
SETI Program Office
Ames Research Center

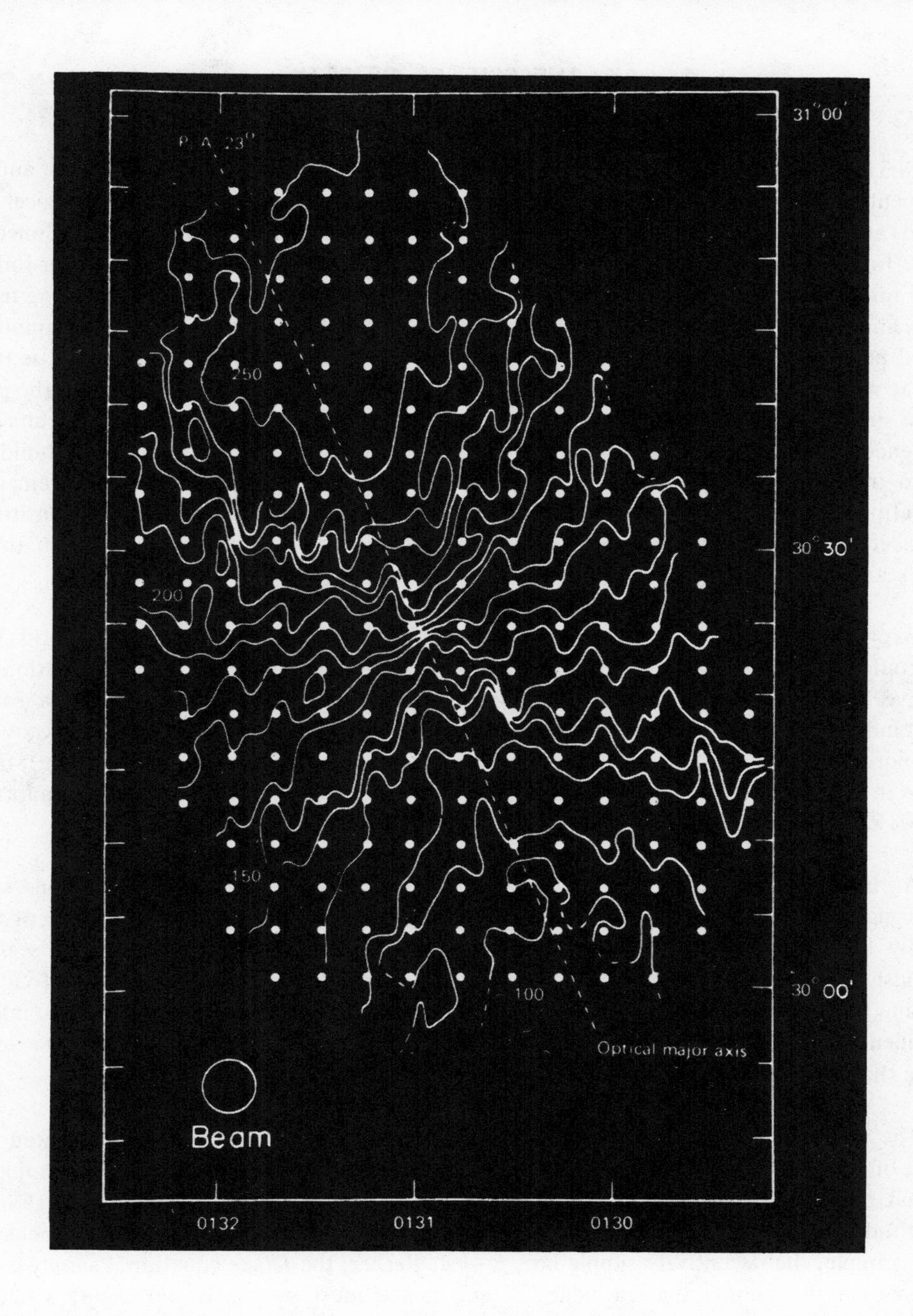

Locations within the galaxy M33, the Great Spiral in Triangulum, where the beam of the Arecibo Telescope has been pointed while searching for ETI signals from any civilizations in that galaxy. The points are superimposed on a map giving the general velocities of stars and interstellar gas throughout M33. With this distribution of points and the telescope beamwidth, every star in the galaxy falls within the coverage achieved with the Arecibo Telescope. Each location was searched for signals for a minimum of 60 sec. At any given instant, about one billion stars were within the beam of the telescope.

6. THE SCIENCE OF SETI

SETI is a manifestation of man's drive to explore. This drive is one of the oldest and most fundamental aspects of our nature; the very origin of the hominidae as a distinct biological entity is owed, at least in part, to the boldness of our venturesome simian ancestors who abandoned their familiar forest environment to probe the savannah, there to seek fleet-footed prey. Our forebears pushed into almost every corner of the globe. They explored by climbing hills, by walking through forests, and even by crossing large bodies of water. Sometimes they may have had in mind some material purpose, but certainly they sometimes went where they did for no other purpose than to see what was there. Modern man still explores, but the arena for his exploration is now the planets and the stars beyond. But we are not limited to physical exploration. We can use the fruits of our intelligence to conduct exploration from a distance. Though some day we may wish to build space ships to travel there, to probe the stars now we need only telescopes. Yet the excitement and exhilaration that comes with this kind of exploration, as larger telescopes and more sensitive and sophisticated data acquisition techniques lead to discovery after discovery, is akin to that the Viking seamen must have felt.

Exploration has always required knowledge and understanding of the physical world. Viking boats could not have been built without knowledge of what wood to use; Viking navigation could not have been accomplished without an understanding of the winds and tides. We call this understanding of nature, which we gain from observation and experiment, scientific knowledge. To explore the stars in search of other intelligent life also requires scientific knowledge; indeed, because it can only be done using highly sophisticated technologies and methods, it requires more scientific knowledge than have man's classical explorations.

A SETI program should embrace not only a search for evidence of other civilizations, such as radio signals, but also a wide range of related scientific studies. We need knowledge of nature primarily for two purposes. One of these is to enable us to narrow the scope of the search by distinguishing promising volumes of search space; for example, we might be better able to identify promising target stars or frequency bands. The other purpose is to enable us to be able to interpret any evidence of other civilizations we obtain, and to decide what course we should follow once we are sure that other intelligent life has been discovered.

The scientific knowledge needed by a SETI program can perhaps best be illustrated in the context of the Drake equation, which relates the expected number N of intelligent, technologically advanced communicative species in the Galaxy to the product of several factors. It should be understood that the Drake equation is not a fundamental expression of the way nature behaves, as is, for example, the deceptively simple law, $f = ma$. Rather, the Drake equation is simply a device to enumerate the factors that influence N and hence must be considered in any attempt to estimate this number. One form of the Drake equation (there are several) is

$$N = R_* f_g f_p n_e f_l f_i f_c L$$

where R_* is the average rate of star formation in the Galaxy, f_g is the fraction of stars that are "good" stars in the sense of providing conditions thought to be necessary for life, f_p is the fraction of good stars that have planets, n_e is the number of suitable planets in a typical planetary system, f_l is the fraction of suitable planets on which life starts, f_i is the fraction of life starts that evolve intelligence, f_c is the fraction of intelligent species that enter a communicative phase and L is the mean lifetime of the communicative phase. It is immediately obvious that the factors on the right-hand side of the equation have widely differing character. Some, such as R_* and f_p, involve only knowledge from one basic discipline, astrophysics. Others, such as f_g, n, f_l and to some extent f_i involve questions spanning many disciplines, including astrophysics, prebiotic chemistry and biology. Finally, the terms f_c and L involve considerations not generally found in the natural sciences, but which nevertheless lend themselves to scientific inquiry. The full panorama of the elements in the Drake equation is embodied in the concept of "cosmic evolution" (see Section II-1).

For convenience, we can take the Drake equation as a model and categorize the science of SETI under the headings of physical science, biological science, and social science. We begin with those aspects of SETI involving the physical sciences.

A very important question concerns the way stars are born. At present, our understanding of this stage in the evolution of matter is sketchy at best. We are able to calculate the mean rate of star formation over the age of the Galaxy (this number is generally used for R_* in the Drake equation), but we are not yet able to specify in detail the conditions necessary for star formation nor are we able to describe or predict the course of events involved in the evolutionary path from a relatively low density interstellar cloud to an incandescent ball of gas supplying its own energy by nuclear fusion deep in its interior. Observational progress on this question has been forthcoming over recent years through infrared and radio studies of dark clouds and newly formed stars; however, many large gaps still exist in the fabric of our knowledge. Theoretical study of star formation is in an even more rudimentary state than is its observational counterpart, due principally to the intrinsically three-dimensional, non-analytic nature of the mathematical representation of the problem, and the inability to model or parameterize the role of turbulence and plasma dynamics in dense interstellar clouds. The advent of very rapid computers is making it possible now to undertake the formidable numerical problems involved in simulating the physics of star formation, and with adequate support we soon should be in a position to compare meaningfully theory and observation. Understanding this fundamental process of star formation is important not only because it provides the means for obtaining a more accurate value of R_* in Drake's equation, but more significantly, it will provide the required construct within which we can understand the formation of planetary systems. An understanding of the formation of planetary systems could allow us to eliminate certain classes of stars as targets of the search.

Some stars, even though they have planetary systems, may be intrinsically hostile to life. There is some evidence, for example, that M-dwarf stars produce powerful flares with accompanying radiation levels inimical to living organisms. There is also the possibility that, as a consequence of the low luminosity of M-dwarfs, a planet in the stellar ecoshell would have to be so close to the star that its axial rotation period would become tidally locked to its orbital period about the star.

However, since M-dwarfs are the most abundant stars, studies which will resolve these questions are important.

One of the primary scientific reasons for the growing feeling that a SETI program is a sound undertaking is the present consensus, based on the previously discussed fragmentary state of knowledge concerning star formation, that conditions conducive to the formation of planets are natural consequences of the birth of a star. We have learned much of the early history of the solar system through our space program. Consequently, we now have a clearer idea how the planets in the solar system formed and evolved. However, as with our understanding of star formation, much work needs to be done. Some of the important questions involve the formation and evolution of planetary atmospheres, and their interaction with their parent star. Do planets form and evolve atmospheres after their star has settled onto the main sequence, do they form and evolve before the star is fully developed, or do both form and evolve simultaneously?

An understanding of this sequence, as well as the details of planetary evolution as a function of mass, composition and its interaction with its environment is important, not only from the standpoint of subsequent biological evolution, but from the standpoints of detecting planets around other stars and identifying those that may have environments favorable to life. Jupiter and Earth emit intense, nonthermal radio bursts, while Venus emits nonthermal, perhaps masering, infrared molecular lines. In spite of their intensity, neither of those characteristic planetary emissions would be detectable over interstellar distances with present technology. Do these phenomena occur on other planets around other stars with sufficient intensity to be detected? If so, as noted in Section II-3, they could provide the best means for carrying out the detection program. Maser radiation from water or some organic compound, or nonthermal radiation arising in a planetary magnetic field might also provide a clue concerning the possible existence of life. We must seek to understand in more detail the nature of planetary formation, the chemistry of planetary atmospheres, and the interaction between star and planet.

Other scientific needs of a SETI program in general, and detection of other planetary systems in particular, involve questions concerning the frequency of binary stars and the stability of stellar properties. As noted in Section II-3, the fraction of main sequence stars that are involved in binary or multiple star-systems, particularly those of short period (periods $\leqslant$100 years), is important. Between two moderately close stars there would be no stable planetary orbits except those with either very short or very long orbital periods. For planets with short orbital periods it is inferred that the planetary surface would be strongly heated by radiation from its parent star. Exactly how hot the surface would be depends on the spectral type of the star and the details of the planet's orbit. For planets with long orbital periods, the opposite temperature extreme might be expected. For this reason, it is important to extend present binary frequency studies to cover a broad spectral range (the most complete study to date falls in the narrow spectral type range F5–G2). If it should turn out that most stars have short period binary companions, we would be forced to conclude that the solar system represents a rare phenomenon in the galaxy, and correspondingly lower the value of f_g. However, such a finding would allow us to concentrate the search on a smaller number of targets.

In connection with the stability of intrinsic stellar properties, it would be useful to be able to determine the long-term stability of stellar luminosities and radial velocities. It is somewhat surprising that we do not know very much about the stability of the apparent solar radial velocity for periods much in excess of a day, let alone months or years. If as discussed in Section II-3, radial velocity techniques are to be considered as potential candidates to be used in search for other planetary systems, we must have a basic understanding of the long-term (comparable to planetary orbital periods) stability of stellar radial velocity as a function of stellar type. This knowledge might also provide needed insight into the modes of internal motion of stars. The subject of stability of stellar luminosities is central to many of the biological aspects of the SETI program. Small scale variations in the luminosity of a star may not be detrimental (and could be beneficial) to the evolution of primitive life forms toward more advanced biological systems, but at present we possess no hard empirical data as to the magnitude or time scale of luminosity variations for any main sequence star, the Sun included. Advances in understanding of the luminosity stability of main sequence stars could also shed light on many fundamental questions related to stellar structure. Arguments have been advanced that the Sun's luminosity varies periodically in time, and that this variation can explain the current lack of neutrinos emitted by the Sun. We simply do not know whether the Sun's luminosity has been, is, or will be varying in time. A potential SETI search strategy could hinge on whether certain spectral types of main sequence stars have sufficiently stable or unstable luminosity to provide conditions conducive to biological evolution.

It will be necessary to develop efficient techniques to identify weak signals embedded in noise, to distinguish signals of artificial origin from natural phenomena, and to evaluate whatever information they might contain. These require studies of pattern recognition at low signal-to-noise ratios, and studies of decoding strategies.

In the biological and social science areas, there is a wide range of studies that are important to a SETI program. In contrast to the purely physical sciences, biological and social science studies have potential not only to contribute to the refinement of search strategy, but also to provide us with guidance with respect to actions that should be taken if another civilization is ultimately discovered. Moreover, the specific problems which SETI would need to have addressed would require new approaches that integrate data and methods from the physical sciences, and these could prove extremely stimulating.

Research is needed on the origin of life (see Section II.1). We now have a general understanding of the chemical evolution process. However, it is not clear in what ways this process depends upon the environment. Both the direction taken by chemical evolution and the time period required for it could be influenced by large numbers of environmental factors including the chemical composition of the planet's atmosphere, the existence and nature of a dense fluid medium, tides, tectonic activity, the degree of climatic diversity, the planetary rotation rate and orientation of the spin axis, the planetary magnetic field, and the energy spectrum of electromagnetic radiation and particles from the primary star. A large number of studies, including laboratory experiments, observations of chemical evolution on other planets in our own solar system, and theoretical studies, can shed light on the degree to which these and other environmental factors influence chemical evolution. Should chemical evolution prove to be very sensitive to a small number of factors, then we will be able to develop programs to allow us to identify systems likely to contain one or more planets upon which those factors exist. Searches for evidence of intelligent life could then concentrate on these systems and ignore others.

The same considerations apply to studies of biological evolution. There probably are planet- or star-dependent environmental factors to which both its course and its rate will be sensitive. Identification of such factors will allow development of a more discriminating search strategy. Such identification can and should be attempted through a variety of chemical, biological and paleontological studies, both in the laboratory and in the field. However, it is important to develop a general theoretical model of biological evolution in the context of cosmic evolution (see Section II-1). The formalism of statistical thermodynamics or other areas of physics that deal with interacting many-body systems might provide useful insight here. Interactions between investigators in many of the physical and biological sciences will be required to develop such models, and such interactions should prove stimulating to all concerned.

But this is not the only contribution studies in the biological sciences can make to SETI. If the search should prove successful, we will face the problem of communicating or otherwise interacting with whatever civilization we discover. In this regard, studies in the area of the social sciences will also be important.

The history of the Earth advises us that interacting with another civilization will be no trivial problem. Members of one culture have often enough failed to appreciate the customs and beliefs of another, and this has led to unfortunate results. It is an observationally determined fact that men of equal intelligence but different cultural backgrounds may not understand one another. Given this, it is not clear that we will be able to decode all the information content of the first signals we receive. Certainly, there is no reason other than faith to believe that, just because both they and we are intelligent, communications that may ultimately take place between us will necessarily convey the intended meaning. To the degree cultural characteristics are determined by biological and evolutionary factors, so may we expect cultural differences between human and extraterrestrial societies to be greater than those observed among human societies. Great cultural differences imply greater potential for misunderstanding. The result of a misunderstanding could range from our failure to comprehend information intended to benefit us, to cessation of transmission.

Human relations, however, also advise us that if we take the trouble to be aware of cultural differences and treat them with respect, relationships can proceed in harmony. Thus our problem is to anticipate how intelligent extraterrestrial societies might be culturally different from us and to determine what communications problems these cultural differences might cause. This is not entirely a problem in metaphysical divination, but is susceptible to at least partial solution by scientific experiment and observation.

In the first place, we have before us the empirical evidence provided by the human cultures of Earth. It should be possible by appropriate studies to determine the factors responsible for their diversity; indeed, much of this work has undoubtedly been done, and needs no more than discussion and synthesis. Models of cultural development which focus on the interaction of a society with its technology could be especially valuable since radio technology is the one characteristic which we believe we will share with any extraterrestrial civilization we are likely to discover. If statistical thermodynamics can supply a useful formalism for models of biological evolution, it may also prove adaptable to describing culture and its development. Moreover, we are not limited to studies of human cultures; evidence is also provided by animal societies. Little has been done to evaluate this evidence, but it may give us some very great insights into cultural evolution, social behavior, and the meaning of intelligence.

The Cultural Evolution Workshop (see Section II-2) emphasized that research since 1960 into animal behavior has completely altered our perception of these creatures. The Workshop members had no difficulty accepting the ideas that man is not the only intelligent animal on this planet; that chimpanzees, small cetaceans, and perhaps elephants should also be included in this category, and that these other species have complex social organizations and systems of communication. Indeed, our similarity to the end products of several other evolutionary lines was cited as one of the principal supports for the proposition that intelligence is likely to evolve on other planets, and is not just a fluke which occurred only here. Investigation of and attempts at communication with other intelligent terrestrial animals might establish the extent to which their "cultures" differ, both intraspecifically and interspecifically, and what factors – biological, evolutionary, or social – are responsible for these differences. The general theoretical model of biological evolution needed for the identification of factors that produce intelligent life will also be important here. A thorough comparative understanding of behavior, encompassing at least men, chimpanzees, and dolphins might provide us at least a minimum of insight into what range of cultural behaviors we might expect among extraterrestrial societies. In particular, comparison of dolphins and primates might give us some feeling for the consequences of different primary means of sensory perception.

The investigation of the nature and origins of behavioral diversity among the intelligent and semi-intelligent higher animals, although not considered to directly affect the present ETI search strategy, nonetheless is considered to be an important parallel effort. Our goals should be (1) to catalog and classify behavioral patterns and cultural differences; (2) to determine how these are related to the environment, physiology, and evolutionary history of each species studied; (3) to determine what traits, if any, appear common to all intelligent animals; (4) to gain experience in actual communication with other terrestrial species; and (5) to develop theoretical models that will allow extrapolation to extraterrestrial cultures, and allow us to evaluate at least semi-quantitatively the uncertainties in such an extrapolation. To the extent that this approach might enable us to better understand human behavior, it could result in one of the most important benefits of the SETI program.

There is little doubt that the science germane to SETI would contribute in a significant way to our overall understanding of cosmic evolution. Positive feedback exists between these studies, such as a search for other planetary systems and analyses of cultural evolution, and a SETI effort. Each can be carried forward independently, but we should make every effort to apply the gains in knowledge from one study to the others.

Prepared by: David C. Black and Mark A. Stull
SETI Program Office
Ames Research Center

SECTION III: COMPLEMENTARY DOCUMENTS

1. ALTERNATE METHODS OF COMMUNICATION*

Several methods of achieving contact with intelligent life beyond our solar system have been proposed. These include actual interstellar space travel, the dispatching of interstellar space probes, and the sending or detection of signals of some form. Many other suggestions involving as yet unknown physical principles have also been made but are not considered here.

INTERSTELLAR TRAVEL

The classical method of interstellar contact in the annals of science fiction is the spaceship. With our development of spacecraft capable of interplanetary missions, it is perhaps not amiss to point out how far we still are with our present technology from achieving practical interstellar space flight, and indeed how costly such travel is in terms of energy expenditure even with a more advanced technology.

Chemically powered rockets fall orders of magnitude short of being able to provide practical interstellar space flight. A vehicle launched at midnight from a space station orbiting the Earth in an easterly direction and having enough impulse to add 10.5 mi/sec to its initial velocity would escape Earth with a total orbital speed around the Sun of 31.5 mi/sec. This would enable the vehicle to escape the solar system with a residual outward velocity of 18 mi/sec, or about 10^{-4} c. One could also use Jupiter in a swingby gravity assist maneuver to escape the solar system (as Pioneer 10 has done) and achieve about the same outward velocity with a somewhat lower launch energy requirement. Since, however, the nearest star, α-Centauri, is 4 light-years away, the rendezvous, if all went well, would take place in 40,000 years. Clearly we must have at least a thousandfold increase in speed to consider such a trip and this means some radically new form of propulsion.

Spencer and Jaffe (ref. 2) have analyzed the performance attainable from nuclear-powered rockets using (1) uranium fission in which a fraction $\epsilon = 7 \times 10^{-4}$ of the mass is converted to energy and (2) deuterium fusion for which $\epsilon = 4 \times 10^{-3}$. The mass ratios, μ, required for a two-way trip with deceleration at the destination are given in table 1 for various ratios of the ship velocity v to the velocity of light c. The mass ratio, $\mu = M_i/M_o$, where M_i is the total initial mass of the rocket (including fuel) and M_o is the mass after burnout, i.e., M_i minus the mass of the fuel. From these figures we would conclude that with controlled fusion we might make the trip to α-Centauri and back in 80 years, but that significantly shorter times are out of the question with presently known nuclear power sources.

Let us ignore all limitations of present day technology and consider the performance of the best rocket that can be built according to known physical law. This is the photon rocket, which

---

*Most of the information presented here has been taken from the Cyclops report, pp. 33–36 (ref. 1).

TABLE 1.– MASS RATIOS FOR TWO-WAY TRIP TO α-CENTAURI

v/c	μ for uranium fission	μ for deuterium fusion
0.1	3.8×10^4	8.1×10^1
.2	2.3×10^9	6.2×10^3
.3		1.1×10^6
.4		1.5×10^8

annihilates matter and antimatter converting the energy into pure retrodirected radiation. The mass ratio μ required in such a rocket is:

$$\mu = \sqrt{\frac{1 + v/c}{1 - v/c}} = \frac{v_{\text{eff}}}{c} + \sqrt{1 + \left(\frac{v_{\text{eff}}}{c}\right)^2} \tag{1}$$

where

$$v_{\text{eff}} = \frac{v}{\sqrt{1 - v^2/c^2}} = \text{coordinate distance travelled per unit of ship's proper time}$$

If we choose $v_{eff}/c = 1$, then to reach the α-Centauri system, explore it, and return would take at least 10 years' ship time. It is hard to imagine a vehicle weighing much less than 1000 tons that would provide the drive, control, power, communications and life support systems adequate for a crew of 12 for a decade. To accelerate at the start and decelerate at the destination requires a mass ratio of μ^2, and to repeat this process on the return trip (assuming no nuclear refueling) requires an initial mass ratio of μ^4. For $v_{eff}/c = 1$, $\mu = 1 + \sqrt{2}$ and $\mu^4 \approx 34$.

Thus the take-off weight would be 34,000 tons, and 33,000 tons would be annihilated enroute producing an energy release of 3×10^{24} J. At 0.1 cent per kWh this represents \$1 million billion worth of nuclear fuel. To discover life we might have to make many thousands of such sorties.

Even disregarding the cost of nuclear fuel, there are other formidable problems. If the energy were released uniformly throughout the trip, the power would be 10^{16} W. But the ship is μ^3 times as heavy for the first acceleration as for the last and the acceleration periods are a small fraction of the total time; hence, the initial power would be about two orders of magnitude greater, or 10^{18} W. If only one part in 10^6 of this power were absorbed by the ship the heat flux would be 10^{12} W. A million megawatts of cooling in space requires about 1000 square miles of surface, if the radiating surface is at room temperature. And, of course, there is the problem of interstellar dust, each grain of which becomes a miniature atomic bomb when intercepted at nearly optic velocity.

We might elect to drop v_{eff}/c to 0.1 and allow 82 years for the trip. But this would undoubtedly require a larger payload, perhaps a 10,000 ton ship, so our figures are not changed enough to make them attractive.

Ships propelled by reflecting powerful Earth-based laser beams have been proposed, but these decrease the energy required only by the mass ratio of the rocket they replace, and they require cooperation at the destination to slow them down. In addition, the laser powers and the mirror sizes required for efficient transmission are fantastically large.

Bussard (ref. 3) has proposed an ingenious spaceship that would use interstellar hydrogen both as fuel and propellant. After being accelerated by some other means to a high initial velocity, the ship scoops up the interstellar gas and ionizes and compresses it to the point where proton-proton fusion can be sustained, thereby providing energy for a higher velocity exhaust. Essentially the ship is an interstellar fusion powered ramjet. Bussard's calculations show that, for a 1-g acceleration, a 1000-ton ship would require about 10^4 km^2 of frontal collecting area. No suggestions are given as to how a 60-mile diameter scoop might be constructed with a mass of less than 1000 tons. (10^4 km^2 of 1 mil mylar weighs about 250,000 tons.) Solutions to this and other engineering details must await a technology far more advanced than ours. For example, it has been suggested that an interstellar ramjet might work by pre-ionization and magnetic scooping using superconducting flux pumps (Sagan, ref. 8).

A sober appraisal of all the methods so far proposed forces one to the conclusion that manned interstellar flight is out of the question not only for the present but for an indefinitely long time in the future. It is not a physical impossibility but it is an economic impossibility at the present time. Some unforeseeable breakthroughs must occur before man can physically travel to the stars.

INTERSTELLAR PROBES

Bracewell (ref. 4) has suggested that advanced societies might build interstellar probes, possibly artificially intelligent and self-reproducing, that expand through the space about their planet of origin, patrolling all star systems likely to develop intelligent life. In this scenario a probe enters a system and may detect stray radio emissions similar to those of our own. Bracewell's postulated response for the probe is to first record signals and then retransmit them to the planet of their origin where the time delay could be interpreted as an indication of the probe's presence and location. Communication with the home civilization then begins via the probe.

Bracewell also suggests we be alert for such probes in our own solar system. Villard (ref. 5) has suggested that long delayed echoes, which are in fact occasionally heard, conceivably could originate from such a probe. Until the source of such echoes can be definitely ascribed to some other mechanism such as slow propagation in the ionosphere near the plasma cutoff frequency, this will continue to be an intriguing, albeit an unlikely, possibility. The phenomenon deserves further study.

An interstellar monitor probe could be much smaller than a spaceship and could take longer in flight. But although there would be no crew to face the psychological barriers or physiological problems of generations spent in space, there are still good reasons to require a short transit time. If the probe were to require 1000 years (or even only a century) to reach its destination, serious doubt would exist that it would not be obsolete before arrival. Thus, even if probes should be capable of velocities of the order of that of light, probe weights in excess of a ton would almost certainly be needed.

To "bug" all the sun-like stars within 1000 light-years would require about 10^6 probes. If we launched one a day this would take about 3000 years and an overall expenditure well over \$10 trillion. Interstellar probes are appealing as long as someone else sends them, but not when we face the task ourselves.

The simple fact is that it will be enormously expensive, even with any technological advance we can realistically forecast, to send sizable masses of matter over interstellar distances in large numbers. This is not to say that unmanned probes to one (or a few) nearby stars for the purpose of scientific exploration would not be a worthwhile endeavor; in all likelihood such an attempt will be made at some future date.

SERENDIPITOUS CONTACT

We cannot rule out the possibility that we might stumble onto some evidence of extraterrestrial intelligence while engaged in traditional archeological or astronomical research, but we feel that the probability of this happening is extremely small. Not everyone shares this view. Dyson (refs. 6, 7), for example, has suggested supercivilizations of various sorts whose activities can be detected even if they are not actively engaged in an effort to contact and communicate with other societies. Dyson argues that a few societies may be unable to avoid a Malthusian population rise and so are forced to redesign their planetary system, then the nearby star system, and finally the entire galaxy in order to gain increased living space and energy resources. In particular, he has imagined a society that disassembles its planets and rebuilds them into a sphere of structures orbiting the central star. These structures are so numerous and so densely distributed that they effectively capture all the stellar radiation and reradiate it in the infrared region of the electromagnetic spectrum. From afar, such civilization would be a strong infrared source that showed other peculiar properties, such as radio or laser emission.

While there is a lack of general support for these and even more imaginative suggestions, some further consideration should be given such scenarios. Indeed, their implicit advantage is that each carries little cost or planning obligation – one only has to go about his normal business and wait either to stumble across evidence of supercivilizations or to be discovered by them. This is an obviously passive approach as compared to the active approach following logically from the orthodox view.

Although no one can deny the excitement that would accompany a physical visit to another inhabited world, most of the real benefit from such a visit would result from communication alone. Morrison has estimated that all we know about ancient Greece is less than 10^{10} bits of information; a quantity he suggests be named the "Hellas." Our problem therefore is to send to, and to receive from, other cultures not tons of metal but something on the order of 100 Hellades of information. This is a vastly less expensive undertaking.

Fundamentally, to communicate we must transmit and receive energy or matter or both in a succession of amounts or types of combinations that represent symbols, which either individually or in combination with one another have meaning – that is, can be associated with concepts, objects, or events of the sender's world. In one of the simplest, most basic, types of communication the sender transmits a series of symbols, each selected from one of two types. One symbol can be the presence of a signal, the other can be the absence of a signal, for example. This type of communication is called an asymmetric binary channel. For the receiver to be able to receive the message, or indeed, to detect its existence, the amount of energy or number of particles received when the signal is present must exceed the natural background. Suppose, for example, we knew how to generate copious quantities of neutrinos and to beam them. And suppose we could capture them efficiently in a receiver. Then, with the signal present, our receiver would have to show a statistically significant higher count than with no signal.

Even if the natural background count were zero, the probability of receiving no particles when the signal is in fact present should be small. Since the arrivals during a signal-on period are Poisson-distributed, the expectation must be several particles per on-symbol. Thus to conserve transmitter power we must seek particles having the least energy. The desirable properties of our signaling means are:

1. The energy per quantum should be minimized, other things being equal.

2. The velocity should be as high as possible.

3. The particles should be easy to generate, launch, and capture.

4. The particles should not be appreciably absorbed or deflected by the interstellar medium.

Charged particles are deflected by magnetic fields and absorbed by matter in space. Of all known particles, photons are the fastest, the easiest to generate in large numbers, and the easiest to focus and capture. Low-frequency photons are affected very little by the interstellar medium, and their energy is very small compared with all other bullets. The total energy of a photon in the microwave region of the electromagnetic spectrum is one ten billionth the kinetic energy of an electron travelling at half the speed of light. Almost certainly electromagnetic waves of some frequency are the best means of interstellar communication – and our only hope at the present time (see Section II-4).

REFERENCES

1. Oliver, B. M.; and Billingham, J.: Project Cyclops, A Design Study of a System for Detecting Extraterrestrial Intelligent Life. NASA CR-114 445, 1972.

2. Spencer, D. F.; and Jaffe, L. D.: Feasibility of Interstellar Travel. NASA TR32-233, 1962.

3. Bussard, R. W.: Galactic Matter and Interstellar Flight. Astronautica Acta, vol. VI, Fasc. 4, 1960.

4. Bracewell, R. N.: Communications from Superior Galactic Communities. Nature, vol. 186, no. 4726, May 28, 1960, pp. 670-671.

5. Villard, O. G., Jr.; Fraser-Smith, A. F.; and Cassan, R. T.: LDE's, Hoaxes, and the Cosmic Repeater Hypothesis, QST, May 1971, LV, 5, pp. 54-58.

6. Marshak, R. E., ed.: Perspectives in Modern Physics, John Wiler & Sons, 1966, p. 641.

7. Cameron, A. G. W., ed.: Interstellar Communications, W. A. Benjamin, Inc., New York, 1963, pp. 111-114.

8. Sagan, C.: Direct Contact Among Galactic Civilizations by Relativistic Interstellar Space Flight, Planetary and Space Science. Science, vol. 11, 1963, pp. 485-498.

Prepared by: John H. Wolfe
SETI Program Office
Ames Research Center

2. NOTES ON SEARCH SPACE

Here we describe briefly some of the major dimensions of microwave search space, noting along the way a few physical and technological factors.

FLUX LEVEL

1. A transmitter with an equivalent isotropic radiated power (EIRP)[1] of (W) watts will provide a flux level of (S) W/m^2 at a receiving site (r) light years away according to the expression

$$S = 8.89 \times 10^{-34}\ (\text{EIRP})\ r^{-2} \qquad (\text{W/m}^2) \tag{1}$$

2. A receiver with a predetection noise power bandwidth (B_r) hertz, an equivalent input system noise temperature (T_s) in kelvins (K), and equipped with a power detector at its output, will provide unity signal-to-noise ratio (*SNR* = 1) after detection and in a time ($\tau \simeq B_r^{-1}$) seconds, if the input power is

$$P_r = kT_sB_r = 1.38 \times 10^{-23}\, T_sB_r \qquad (\text{W}) \tag{2}$$

In the microwave region it is practical to choose bandwidths in the range $10^{-5} < B_r < 3 \times 10^8$ without appreciably affecting the system noise temperature. A T_s of 10 K or less is achievable on Earth or in space. An ideal microwave receiving system in free space would have a T_s of at least 2.7 K because of the cosmic background radiation.

For optimum *SNR* with a signal of intrinsic bandwidth B_t,

$$B_r \cong B_t \tag{3}$$

Of course, by averaging n observations of duration τ, the *SNR* can be improved by $(n)^{1/2}$, for $n >> 1$. However, any modulating information of bandwidth $B_m > B_r/n$, will be lost.

We choose a nominal B_r = 0.1 Hz for this discussion, and T_s = 10 K. Therefore,

$$P_r = 1.38 \times 10^{-23} \qquad (\text{W}) = -228.6\ (\text{dBW}) \tag{4}$$

When T_s and B_r are chosen, the gap between equations (1) and (2) can be overcome at the source through choice of P_t and g_t. At the receiving end the signal collecting area is the one remaining variable at choice. (The use of n unit observations, or post detection integration, is

---

[1] EIRP = P_tg_t, the product of transmitted power and directive antenna gain in the direction of the receiver. For an isotropic radiator, g_t = 1.

reserved here for the explicit purpose of providing an adequate false alarm probability. The single objective is to detect the presence of a signal. Decoding any modulation which may be present, is another matter.)

3. For $S/N = 1$ and effective collecting area A_e in square meters, the range equation is

$$r = [A_e(\mathrm{EIRP})/4\pi k T_s B_r]^{1/2} \quad \text{(m)} \tag{5}$$

or

$$r = 20[n_{100}(\mathrm{EIRP}-9)]^{1/2} \quad \text{(ly)} \tag{6}$$

where, n_{100} is the number of 80 percent efficient 100-m dishes and the unit of (EIRP-9) is a gigawatt (10^9 W). Equation (6) is plotted in Section II-5, figure 2.

4. One can hope that the transmitting society uses large values of (EIRP).

5. The receiving antenna area is likely to be the single most expensive capital item in a large, passive search system. The 1.5 GHz antenna gains of a few sizes of radio telescope are given in table 1 below, and were calculated assuming 80 percent efficiency ($A_e = \eta A_{geom}$). A 90 percent efficiency should be available. The full half-power beamwidths of these antennas are also indicated in table 1.

TABLE 1.– TYPICAL ANTENNA GAINS, EFFECTIVE AREAS, AND FULL HALF-POWER BEAMWIDTHS AT 1.5 GHz

Telescope diam	g_r	A_e	$\theta_{1/2}$
25 m	1.23×10^5 = 50.9 dB	3.13×10^2 m^2	29′
100 m	1.97×10^6 = 62.9 dB	5.03×10^3 m^2	7′.2
213 m	8.96×10^6 = 69.5 dB	2.28×10^4 m^2	3′.4

$$g_r = \eta(4\pi A/\lambda^2) = \eta(\pi d/\lambda)^2$$

$$\theta_{1/2} \simeq 3600\,(\lambda/d)\ \text{arcmin}$$

6. A reasonable expectation for the ground-based, zenith noise budget of T_s, assuming a low-noise antenna at a good site, is given in table 2. A space-based system would total at least 2 K less. In and near the galactic plane, or when a discrete radio source is in the beam, T_s will increase appreciably, and particularly so with high values for the gain (g_r).[2] The atmospheric contribution varies approximately as the secant of the zenith angle.

[2]In the case of an antenna array, g is the gain of the array element.

TABLE 2.– ORIGIN OF THE SYSTEM NOISE TEMPERATURE WITH THE ANTENNA POINTING AT THE ZENITH FROM A QUIET SITE AT ~2000 m ALTITUDE

Galactic background (minimum)	~4 K
Atmospheric radiation	~2 K
Maser input amplifier	~2 K
Antenna noise due to imperfections	~2 K
T_s	~10 K

7. Bandwidth and effective area are inversely related for constant range (r). Clearly one must chiefly count on carrier detection (unless a stunningly powerful signal has been overlooked). That is, the type of signal which is easiest to *just* detect, is one possessing a high spectral density. For instance, consider UHF TV. About one-fourth of the transmitted peak power is in the steady carrier component, which usually has a bandwidth much less than 1 Hz, when observed for some tens of seconds. The remainder of the TV signal power is spread across about 4×10^6 Hz, and is highly variable as well. Because the system noise is proportional to B_r in Equation (2), it is several million times easier to detect the carrier than to detect the modulation.

8. To avoid excessive false alarm signals just due to random background noise peaks, Project Cyclops suggested, in a thorough discussion of detection theory, that about 100 unit observations at $S/N \simeq 1$ should be averaged. This is another way of saying that one must accept a useful limit such that $S/N \simeq 9$ regardless of how it is achieved, via collecting area and/or integration time, when dividing the total instantaneous search band into $\sim 10^9$ channels.

9. To allow for either very long integration times ($n >> 100$) on nearby objects, or more sophisticated signal analysis, one would like a situation where all interfering terrestrial signals produced a receiver input power under −250 dBW. This is 50–100 dB below normal communication technology, and it is why a search system on Earth cannot tolerate satellite transmission in (or too near) the search frequency band.

10. Radio astronomy search limits during normal observing procedures, are discussed in Section II-5. Explicit coherent signal searches using normal radio telescope systems, have covered few directions and few frequencies so far (see Section III-12).

11. A useful antenna figure of merit (AFM), in the present context where total collecting area and operating frequency are unspecified is,

$$AFM = \eta(T_s \$m^2)^{-1} \tag{7}$$

where $\$m^2$ = (total unit antenna dollar cost)/(geometric aperture area in square meters).

SEARCH DIRECTIONS

1. The number of independent pointing directions (n) for an antenna gain g is $n \cong g$. This and related matters (e.g., all sky area search, target search, and the need for good target lists) are discussed in Sections II-5, III-3, and III-4.

2. A single, one-feed, space-based antenna system can essentially observe in any direction in the whole sky at any time, and in one direction continuously, while a single Earth-based system cannot cover the whole sky nor observe in one direction continuously unless the direction is circumpolar. As a practical matter, it would take five or six Earth-based antenna systems to achieve these capabilities of one range-equivalent space system. Of course, the multiple Earth stations could carry out a given all sky area or target search in one-fifth to one-sixth the time. If, as has been suggested, a large space antenna can be equipped with three feeds, this advantage is reduced by a factor of three.

3. If one assumes merely that transmitting species are randomly distributed among, say, F, G, and K dwarf stars, they will tend to share the distribution pattern of these stars. In the neighborhood of the Sun, the density of these stars seems to be uniform in the galactic plane (so far, perhaps, because the data is yet uncertain), but falls to half density at about ±650 ly in directions normal to the plane. Then, if one cannot see the entire sky with a given search system, one might sample a given total number of target stars by increasing slightly the range calculated with equation (6), picking more targets from directions near the galactic plane. On this simple assumption, even a system with moderately limited sky-coverage (spherical reflectors in the ground, for instance, like the Arecibo installation) could test such hypotheses as, "There is a 0.95 likelihood of finding one detectable species among a random assortment of (N), F, G, K dwarf stars." The volume of space surveyable varies as the 3/2 power of the quantity inside the brackets of equations (5) and (6).

4. Both these uniformity assumptions are clearly simplistic. Again, not all F, G, K stars are the same age as the Sun: Nor are they equally rich in the heavier chemical elements. Information such as this is needed in order to optimize search strategy; thus there should be an effort to improve the specificity of the target list and to improve our knowledge of the actual distribution of the targets in direction and range. (This matter is discussed further in Section III-4).

FREQUENCY DOMAIN

Search Bands

The arguments in favor of the free-space microwave window and the water hole in particular, are given in Section II-4. As a practical matter, all search bands must be free of wideband RFI for a significant fraction of the observing time. On Earth this means no line-of-sight transmitters, and some relatively minor protection from over-the-horizon systems. This could be arranged, in all likelihood, for perhaps 25–35 percent of the spectrum between 1 and 10 or 15 GHz. All bands in which visible satellites are transmitting, or high power surveillance radars are operative, are essentially useless for search purposes from even a so-called quiet site. With a space-based search system, essentially no sharing of the search band with either Earth-based or space-based transmitters is possible, unless an RFI shield separate from the antenna is provided.[3] Such a shield should be on the order of two to three times the diameter of the antenna it is protecting, and its edges should be treated to prevent signal currents from propagating on the back surface. So large a diameter is required in order to attenuate, by diffraction loss, interfering signals from the Earth and its vicinity (out to synchronous orbit, at least). The electromagnetic design, physical construction, and cost of such shields needs study.

Instantaneous Bandwidths

Aside from RFI considerations, several technical factors limit the width of the search band that can be simultaneously observed.

1. The cost of a single large collector system drives home the need for antennas with high diffractive efficiency and low dissipative losses, which latter usually cluster in the feed assembly and could, of course, be cooled sufficiently by moderately large cryogenic systems. Besides efficiency, one needs a "low noise antenna." Large antennas today are usually magnified versions of the classical small antennas developed in an era when truly low noise amplifiers were not available. Thus we have, in the main, axisymmetric prime focus or cassegrain systems. None are efficient, i.e., $\eta > 0.85$, nor are they low noise systems when compared to what can be achieved with modern electromagnetic technology.

Above 1 GHz, the dimensions of practical large antennas are also large when measured in wavelengths, but they are still point-focus devices and the aberration problem is simple compared to that in wide field-of-view optical designs. A low noise antenna needs to be an off-axis device, one with no obstructions anywhere in the wave front path. Avoiding wave front blockage improves efficiency and avoids most structural scattering of unwanted signals and thermal noise into the feed from the surroundings. By shaping all mirrors in the Galindo sense,[4] one can achieve both high efficiency and low wide-angle and back lobes – hence a minimal noise contribution from the ground and lower atmosphere.

Optimum feed design needs study. The desirable properties of a feed are low dissipative losses, low spillover past the secondary reflector, and an electromagnetic field geometry, in both intensity and phase, that is nearly independent of frequency and polarization over the instantaneous search band. Furthermore, these properties should remain constant as one changes feeds

---

[3] An exception to this could be, of course, a search system in solar orbit, stationary with respect to the Earth and one Earth-Sun distance (1 AU) or more, away from the Earth.

[4] V. GALINDO, "Design of Dual-reflector Antennas With Arbitrary Phase and Amplitude Distributions," IEEE Transact. Ant. and Propag. AP-12, p. 403, 1964.

when changing search bands. Low noise, efficient feeds in current practice, with the exception of the Hogg Horn, tend to have bandwidths less than ~25 percent.

Shaped antenna systems have a long wavelength cut-off in the sense that for $\lambda > \lambda_c$, the diffractive performance drops from 90 percent or better, down to the 50 percent level. This occurs when the secondary mirror dimensions are no longer many wavelengths in size, or when the feeds and secondary mirrors required for high performance are no longer practical structures. It is fortunate that for the 100 m and up antenna diameter systems, λ_c falls outside the long wave boundary of the microwave window, where ultra low noise feed systems are no longer so critically important.

2. Another limit to the instantaneous system bandwidth is set by ultra low noise amplifier technology. Maser designs tend to have a constant gain bandwidth product, and a fractional bandwidth at a given gain that is independent of center frequency. To cover the water hole requires a 0.21 fractional bandwidth. This is difficult to attain with a single maser, though it appears possible after sufficient research and development.

A more attractive solution would appear to be the development of helium-cooled up-converters to feed a maser with the required bandwidth that operates at some frequency in the 20–30 GHz region. This scheme has the advantage that only two masers (one for each polarization), each with seven to ten up-converters (all in the same cryogenic package), can cover the entire terrestrial microwave window. This assumes, of course, that one can construct up-converters with $T_s < 2$ K. Extrapolating experience and current understanding strongly suggests this is a realistic expectation.

3. The possibility of simultaneously scanning two or three well separated frequency bands should be examined, since it would speed the spectral search accordingly. It would be particularly appropriate for the target search mode if it can be done without appreciable damage to the system noise temperatures. (Multiplying the number of simultaneous search bands means, of course, also multiplying the multichannel capability of the receiving system.) If applied to the area search mode so that at the highest frequency one just completes a nonoverlapping full-sky search in time τ_s, then the lower bands being scanned at the same time would have covered the sky many times over, in proportion to the ratio $(\lambda_L/\lambda_H)^2$, unless one used an interleaved scan strategy with the highest frequency area search and changed the lower frequency scans to adjacent bands whenever a nonoverlapping search had been completed. This strategy would end up with roughly equal flux level limits in each frequency band instead of the $\sim\lambda^{-2}$ relationship when all-sky search times are equal (see Section III-3).

4. Since the chief, initial search mode in the frequency domain is carrier search, we estimate here some dimensions of the spectral data processor. It consists of two main units, plus the scavenging system that selects and compacts the archival data. These two units are the Fourier transform filter-processor and the pattern detection system.

At the time of Project Cyclops (1971) an optical-photographic-magnetic disk design was proposed as the cheapest feasible system. Digital costs have dropped so far since then that a totally digital system is now cheaper than the optical, as well as less costly to operate and more reliable than a mainly analog system.

Because the search range in carrier search is inversely proportional to the square root of the bin width (B_r in eq. (5)) as long as the observed carriers stay within the bin during the time $\tau \cong B_r^{-1}$, we estimate a minimum practical bin width to be perhaps 10^{-2} Hz (see the Cyclops Report, pp. 55–58 (ref. 1) for relevant discussion). We can remove our doppler drift to that accuracy, receiving or transmitting, so we must assume the transmitting species could do as well if they cared to do so. Since they may not, larger bin widths may be needed, as they are for area search procedures and for eavesdropping; or for spotting pulsed signals, or frequency or directionally scanned signals. Thus the multichannel spectrum analyzer should be able to provide bin widths from 10^{-2} Hz up to perhaps several kilohertz.

Now let us estimate the magnitude of the extreme bit memory requirements of such a system, assuming instantaneous coverage of the water hole.

Bin width	10^{-2} Hz
Total bandwidth	3.27×10^8 Hz
Number of unit observations per target	100
Number of polarizations	6
Bits/bins	16

This comes to 3.14×10^{14} bit memory cells which must be scanned for signal patterns and scavenged for archival data every 10^4 sec. This may be the largest digital system seriously envisaged to date, but it is feasible within a decade. Even if the bin width is limited to 1 Hz, reducing the scavenging period to 10^2 sec and the temporary bit memory call to 3×10^{12}, use of such a system with current radio telescopes would improve their carrier search capability by at least 35 dB. In reality, the pattern recognition capability of such a system would improve the "ETI signal identification sensitivity" still further by several orders of magnitude.

At present a 10^6 complex bin, flexible, modular prototype analyzer is in design, and completion of assembly could be expected within two years after funding. It has these properties:

Input bandwidth	$\leqslant 4 \times 10^6$ Hz complex	
Output bins	$2^{20}/2^m$	$m = 0, 1, 2, 3 \ldots$
Bits/bin	16 or less	
Input bits	4 – 8	
Input sampling rate	dual channel and up to 10^7 Hz	
Maximum output field	100×10^6	
Output bit memory cells	3.2×10^9	

This unit is designed to be a test bed for:

1. Acquiring on-air experience and proving (and improving) the economy of the design.

2. A preliminary look with high resolution at a number of interesting areas and objects in the sky.

3. The development of pattern recognition algorithms.

4. Appraising its suitability for other applications.

5. Determining the response of the design to RFI.

This experimental design uses a variety of available chips and microprocessors, and is economical on this scale at this time. It is not clear whether or not this particular design is suitable for extension to a 10^9 bin system, now or in the future.

The size of a 10^9 bin system, the need to retain organizational flexibility so that strategies in data analysis may be changed as experience is gained, and reliability and service needs – all these factors argue firmly for ground-based data processing even if space-based antennas are used.

POLARIZATION DOMAIN

At least two RF channels should be used in each search frequency band, one for each of two orthogonal polarizations. From these two, four additional polarizations may be synthesized at the central processing station to give a total of six: $V, H, V+H$ (or 45°), $V-H$ (or 135°), $V+jH$ (left circular) and $V-jH$ (right circular). If all six are processed, the probable loss for a signal of unknown polarizations, is 0.4 dB or 10 percent, and the maximum loss is 0.7 dB, or 17 percent. If only four polarizations are processed, the probable loss is 0.7 dB and the maximum 3 dB, a factor of 2 and quite equivalent to discarding half the antenna in use. Hence six spectrum analyzers are desirable in a fully built-up system. If fewer processors are available at some stage, one can compensate, assuming constant signals, by sequentially observing with different polarizations, thus trading time for processors. With six processors there may be a small but worthwhile increase in sensitivity if during the 100 unit-target-observations 50 of the unit observations are carried out with the four linear polarizations shifted by 22.5°.

There is no predicting what polarization schemes another species might employ, except (probably) in the case of signals expected to be received at great interstellar distances by antennas of unknown rotational orientation. Magnetoionic plasmas in planetary atmospheres and in interstellar space, particularly in the region of the Galactic plane, will rotate the plane of polarization of a plane polarized wave (the Faraday effect), and for maximum response the polarization of the receiving antenna must equal that of the incident wave. The polarization of a circularly polarized wave, on the other hand, will only be altered under most exceptional circumstances, to the best of our observational knowledge. Furthermore, the response of a circularly polarized antenna of the proper sense is independent of the rotation angle around the bore sight axis. Thus one would expect intentional, long distance signals to be circularly polarized at the point of origin.

MODULATION

Electromagnetic waves may be modulated in any one of these ways – in amplitude, in phase and/or frequency, or in polarization – or in any combination of these. We have used all these degrees of freedom to some extent. It is not absolutely necessary to have strong carrier components present with the modulation sidebands, but carriers or subcarriers, bearing an appreciable fraction of the total power in very narrow bandwidths within the total signal, are the general rule in our technology. The redundancy inherent in carriers simplifies coherent detection of the information in the modulation. Sometimes we suppress carriers somewhat in order to save power, but these vestigial carriers, as they are called, are prominent compared to individual frequencies in the sidebands. In order to make most efficient use of the radio spectrum, we have found it increasingly advantageous to thoroughly stabilize these carriers, hence they have very narrow frequency spectra. At the least, many of our signals are relatively easy to detect, if not to decode.

What a technologically advanced species might find most useful in the complex modulation domain, is unknown. Since there is only one electromagnetic spectrum, our experience would suggest that they may use it efficiently, in the Shannon sense. If they do, carriers may not exist in the prominent fashion to which we are, so far, generally accustomed. If this is the situation, attempts to eavesdrop over a few tens or hundreds of light years will be more difficult by unknown orders of magnitude. What appears patently clear is just this. The basic EM communication physics of our Universe seems to be fairly well understood by us here on Earth. Thus, if we or another species want to make signals detectable at great distance, very stable, high spectral density signals will be provided. Similar signal characteristics are eminently desirable if we are to eavesdrop on unintentional, intraspecies transmissions.

TIME FACTORS

Signals that appear at regular or irregular intervals, perhaps due to spacial scanning, time schedules or frequency programs or schedules, are a form of modulation, of course, but are separately discussed here for convenience of emphasis.

1. A way to produce high flux level signals at great distance is to use a very directive antenna. There are at least two obvious and simple strategies here. Intentionally, such powerful signals may be scanned over a part or the whole of the sky, appearing in a given direction at regular intervals in the form of a strong pulse of some appreciable duration. Or the transmitting species may have determined by means presently beyond us, that there are only a small set of likely directions within (their) reasonable range and they confine their transmissions, simultaneously or sequentially, or irregularly, for short or long sampling periods, to these directions.

The possibilities one can visualize depend on our imagination, but we should be able to rank them in an order of estimated likelihood and be on the lookout for the more distinctive time patterns. At the moment there seems to be no way to avoid this large range of time dimensional possibilities.

2. It has been suggested occasionally by some radio observers who have suffered interference from swept frequency transmissions by ionosounders, chirp radar, and the like, that an ETI beacon might be systematically swept in frequency in order to assist with its identification as an artifact, and to make less demanding the frequency-search aspect of SETI at the receiving end.[5] The apparent attractiveness of this tactic is diminished by recognizing that unless one can derive a singular frequency sweep rate from obvious and likely, universal, physical arguments, it requires increasing the receiver bin width, which increases the signal power required for detection. Then, too, if the receiving frequency-search effort is to be significantly reduced, still more power (costly to us at any rate) must be continuously supplied per hypothesized signal detection possibility, and a minimum observing time interval is established by the frequency sweep repetition rate. To recognize and identify an ETI signal requires more than a single, band limited pulse.

Similar arguments pro and con, can be hypothesized with respect to the possibility that ETI beacon signals might be pulsed in amplitude. So far, only one thing seems clear, and two others rather persuasive, as a result of human experience.

a) A frequency stable signal is much easier to detect than a signal of the same power which is gyrating in time and/or frequency in an unknown way.

b) Unless the transmitting society knows where we are and the likely state of our technology (which they might, of course, if they have observed our radiations), beacon maintenance is a power consuming operation. There, as here now, there may be a need to conserve energy.

c) They may, also, have a strong interest in spectrum conservation because of their own interference problems; and then a stable signal, narrow in frequency, would seem a more likely choice.

3. As yet, we have no generally useful algorithm for recognizing coherent signals when the SNR/Hz is appreciably less than unity, except the simple one of long integration time and precision examination of the relative level of one part of the spectrum compared to the spectrum levels on either side. Again, a frequency stable signal is much easier to detect.

4. Finally, we note that intrinsically stable signals launched uncompensated from rotating planets revolving about a central star do show distinctive, informative, Doppler drift patterns.

---

[5]It is likely that, in fact, it was the intensity of the signal that truly mattered, not its characteristic behavior in the frequency-time domain. The latter merely identified the nature and origin of the signal as the product of a particular, narrow class of services.

PATTERN RECOGNITION

1. The need for pattern recognition in the frequency-time domain is a pervasive theme in search strategy discussions. How should one best seek to recognize the presence of an almost infinite class of coherent patterns in a finite noise field and at low SNR? Can powerful and rapid algorithms be developed to answer this question?

The human eye, ear, and brain are probably our most versatile pattern perceivers, so far. But the human being is fallible, pattern selective, imprecise, subject to fatigue and hallucination, and above all in this context, too costly.

Fortunately, there are algorithms for recognizing simple lines and bands in a frequency-time data field, for recognizing spectral lines and the like. Figure 1 is a famous picture which here simulates a slowly drifting carrier in a noisy frequency-time field. Eye or machine recognition for such a pattern is straightforward. In the figure, imagine frequency increasing to the right, and time increasing downward.

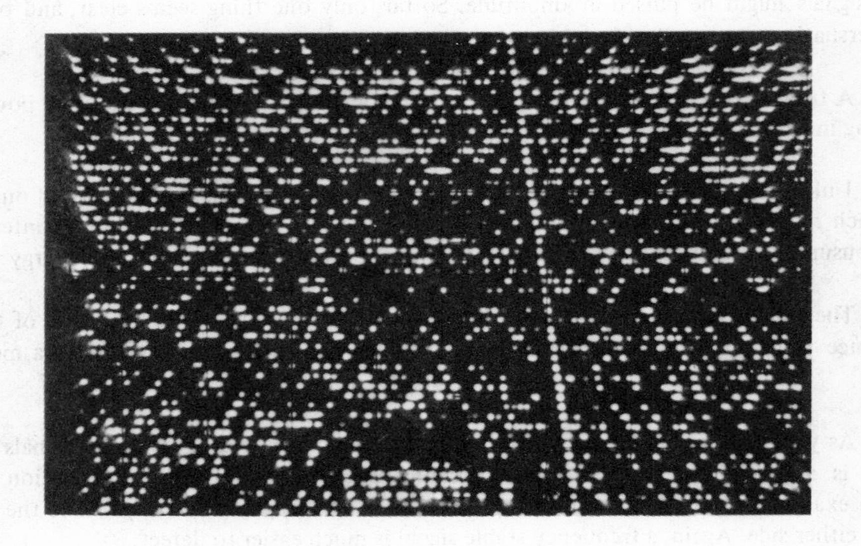

Figure 1.– Signature of a pulsar produced by simultaneous observation on adjacent frequency channels. (Photograph courtesy of Martin Ewing, Calif. Inst. of Technology.)

We do not wish to overstress or understress the importance of pattern recognition studies in the development of search strategies. A major research effort seems worthwhile. But for a significant class of signals *which we think are very likely signals*, we do not have search procedures already defined, and our concern is mainly to find the most efficient procedures.

2. A visual pattern display with pedestal blanking and coordinate compression capability is clearly needed for diagnostic purposes and for pattern recognition studies. It should be possible to

"zoom" a $10^3 \times 10^3$ point display at will over the stored data field, and with adjustable magnification (or compression). Assistance might be sought from those who have been studying visual pattern recognition problems. This visual display should be given an early priority and, at least in prototype form, be used with the earliest high frequency-resolution observations, the better to bound the problem area for the first automatic scanners to be developed.

3. Multicomponent natural spectral line signals should be recognizable and probably fitted to compact descriptions by multicomponent Gaussian approximations. Likewise, it should be possible to recognize pulsing signals, narrow band, broadband, and showing dispersion. Since the search system must be gain stable, the galactic background noise level as a function of frequency, should be recorded with minimal redundancy.

4. There should be algorithms for recognizing carefully defined categories of RFI, coherent or not. Following up false alarms is tedious, subtracting directly from search time, and a practical balance must be struck between sensitivity and false alarm rate.

5. To summarize, the field of pattern recognition is an important and rich one to study.

CHARACTERISTICS OF OPERATING RADARS

Table 3 lists a number of the most powerful radars operating in the territories of the United States and some other nations. It is tantalizing to realize that if another intelligent species should somehow recognize the solar system as a likely site for intelligent life, then it would be trivial to illuminate it with an easily detectable signal from enormous distance.

TABLE 3.– LOCATION, FREQUENCY, AND EIRP OF MOST POWERFUL RADARS IN 1000–2500 MHz BAND

Location	Frequency, MHz	Power (EIRP), W	Number
Arecibo[a]	2380	7.1×10^{12}	1
Goldstone, Calif.[a]	2100	3.2×10^{12}	1
Goldstone, Calif.[a]	2388	7.9×10^{11}	1
Westford, Mass.[a]	1295	1.6×10^{9}	1
Stockholm, Sweden	1315	5.7×10^{9}	1
Argentina (Ezeiza)	1324	2.2×10^{9}	1
Bahrain	1795	2.0×10^{9}	1
Bahrain	1825	2.0×10^{9}	1
United Arab Emirates	2105	2.0×10^{9}	1
United Arab Emirates	2075	2.0×10^{9}	1
Roetgen, Germany	2353	1.3×10^{9}	1
Roetgen, Germany	2395	1.3×10^{9}	1
Goldstone, Calif.	2100	1.2×10^{9}	4
Madrid, Spain	2101.8	1.2×10^{9}	2
Madrid, Spain	2106.4	1.2×10^{9}	2

[a]Experimental, interplanetary radars.

Prepared by: Charles L. Seeger
SETI Program Office
Ames Research Center

3. PARAMETRIC RELATIONS IN A WHOLE SKY SEARCH

It has been argued that in addition to searching likely main sequence stars for signals of intelligent origin, we should also search the entire sky (see Section II-5). While this cannot be done to as low a flux level in the same time, our uncertainty as to the highest flux level we might find is very large and we cannot rule out the possibility that such a search would discover a signal.

We shall derive expressions relating antenna size, sensitivity, bandwidth, and search time in systems that are designed to optimize the cost-performance ratio. We find, not surprisingly, that to conduct full sky searches to flux levels of 10^{-24} to 10^{-26} W/m^2 requires long times and large antennas. However, just as there are good reasons for conducting a targeted search at sensitivities less than that of a full scale Cyclops system, so also there are good reasons to search the full sky with existing antennas at lower sensitivities than we consider here.

To be significant, a full-sky search should be capable of detecting coherent signals at least one or two orders of magnitude weaker than would have been detected by past radio astronomy sky surveys. This is not as difficult as it first seems because radio astronomy surveys have actually discriminated against coherent signals of the type we are seeking and are generally quite restricted in the frequency coverage.

As in the case of a targeted search, we shall assume a monochromatic signal. If there is modulation we ignore it, initially, and try to detect only the strong CW carrier that may be present.

DETECTABILITY OF PULSES

Whether we point the antenna in a succession of sidereally fixed directions until we have tessellated the whole sky, or sweep the beam along some path that paints the entire sky, a CW signal will be received as a pulse embedded in random noise. In the first instance the pulse will have a constant amplitude and a duration, τ, equal to the dwell time per direction. In the second, the pulse envelope will be determined by the antenna beam pattern as it sweeps past the source.

The best possible signal-to-noise ratio (SNR) for the detected pulse is $2W/\Psi$, where W is the received pulse energy and Ψ ($=kT$ in the microwave window) is the noise power spectral density. This optimum can be achieved in several ways, such as with a matched filter and a synchronous detector (Cyclops report, appendix C (ref. 1)). Lacking a priori frequency and phase information we cannot use synchronous detection and can only achieve

$$\mathrm{SNR} = \frac{W}{\Psi} \tag{1}$$

The important point is that the SNR depends only on the pulse energy and not at all upon the way that energy is distributed in time.

NUMBER OF POINTING DIRECTIONS

An antenna that radiated uniformly into a hemisphere would have a gain of 2 and would have to be pointed in two directions to cover the sky. Similarly, one that radiated uniformly into an octant would have a gain of 8 and would require eight directions. In principle, if n is the number of pointing directions and g is the gain we have $n = g$.

However, practical antennas do not radiate uniformly into a certain solid angle; instead, the gain falls off smoothly as an analytic function of the off-axis angle. In the Cyclops report it was asserted that to cover the sky with a maximum off-axis pointing loss of 1 dB at the periphery of each elemental patch of sky would require $n \approx 4g$.

This is true so long as no record is kept of cases that fail to exceed the threshold but are nevertheless strong. If this is done, the record can be used to confirm the presence of a signal as soon as an adjacent pointing direction also shows a strong level at nearly the same frequency.

Assume that an antenna is being pointed successively in a set of directions separated by one half-power full beamwidth and forming a hexagonal lattice as shown in figure 1. Assume that there is a signal at A. When the antenna is pointed at a_i the received signal is 3 dB below threshold, and at a_{i+1} it is again 3 dB below threshold. But suppose we observe that both signals are strong and record them both. If we average the two results and use a threshold appropriate for two independent samples, then we find from figures 11-14 of the Cyclops report that we will have improved our sensitivity by ~2.7 dB. Thus, we will lose only ~0.3 dB for a signal from A as against one from a_i or a_{i+1}. The same is true for a signal from B if the observations a_i and b_i are averaged. For a signal from C we might average the three strong signals from observations a_i, a_{i+1},

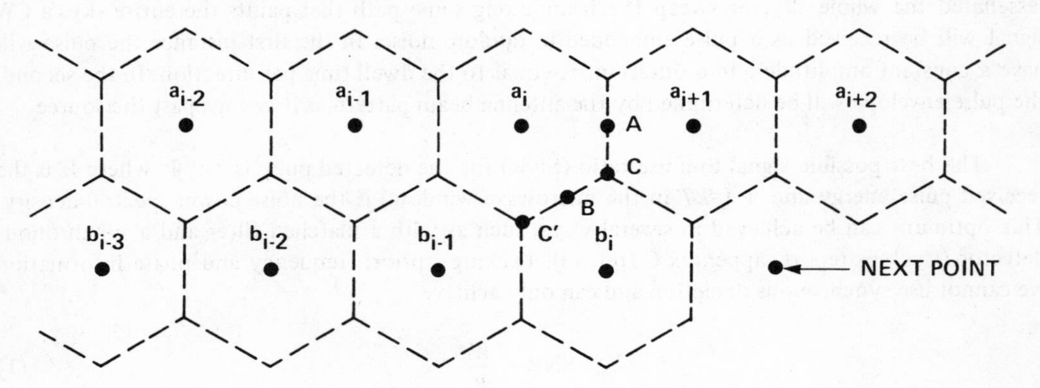

Figure 1.– Off-axis signal detection scheme.

and b_i. Using the appropriate threshold we find from figures 11-14 of the Cyclops report an improvement over one observation of ~4.1 dB and this is exactly the off-axis loss at C. Hence for signals from C we have no loss in detectability as compared with on-axis signals. We could probably do as well at B as at C by including in some way the observations at a_{i+1} and b_{i-1}.

We need to analyze and optimize some set of rules such as:

1. If a sample received at b_i is above a threshold x_1, sound an alarm.

2. If the sample at b_i is less than x_1 but greater than x_3 where $x_3 << x_1$, record it.

3. If the sample at b_i is greater than some threshold x_2, where $x_3 < x_2 < x_1$ and this is also true for a sample at b_{i-1}, a_i or a_{i+1}, add the two and compare against the appropriate 2-observation threshold.

4. If the sample at b_i and two others at b_{i-1} and a_i or a_i and a_{i+1} are all greater than x_3, add and make a 3-observation test.

5. Record b_{i+1} and erase a_i.

The technique of tessellating the sky with an array of fixed pointing direction must be used if the multi-channel spectral analyzer (MCSA) speed limits us to one observation per beamwidth. If, as with an optical processor, continuous spectra can be formed, then the antenna can be scanned continuously. Off-axis signals are simple to detect in this case, since the maximum found on two successive scans will occur at the same value of the coordinate along the scan direction and will be caused by a signal at this same coordinate value.

As shown in the appendix, if the hexagons of figure 1, or the strips of a continuous scan are a full half-power beamwidth wide, then

$$n \approx g \tag{2}$$

SYSTEM SENSITIVITY

If the received flux is ϕ W/m^2 at wavelength λ, and the antenna area is A, the gain will be $g = 4\pi A/\lambda^2$ and the (on axis) received power will be $A\phi$. The effective duration of the received pulse is

$$\tau = \frac{t_s}{n} \tag{3}$$

where t_s is the time required to search the entire sky in one frequency band. We thus find

$$W = A\phi\tau = A\phi\frac{t_s}{n} = \frac{A}{g}\phi t_s = \frac{\lambda^2}{4\pi}\phi t_s \tag{4}$$

The received pulse energy is independent of the antenna area and is the energy that would be received by an isotropic antenna in the full sky search time.

We must set our detection threshold high enough that noise alone rarely exceeds the threshold. This will require a SNR = m where $m \approx 25$. Thus

$$\phi_o = \frac{4\pi\Psi m}{\lambda^2 t_s} \tag{5}$$

or conversely,

$$t_s = \frac{4\pi\Psi m}{\lambda^2 \phi_o} \tag{6}$$

Figure 2 shows ϕ_o as a function of t_s for various λ with $\Psi = kT$, $T = 10$ K and $m = 25$.

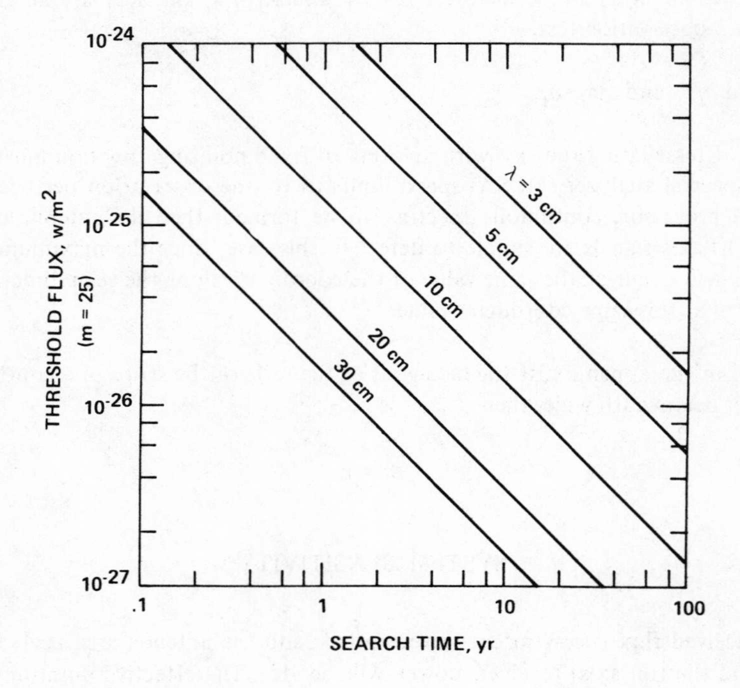

Figure 2.– System sensitivity relations.

FREQUENCY COVERAGE

Although for a fixed search time the sensitivity does not depend upon antenna area, there are reasons to prefer a highly directive receiver. If a signal is discovered we want to know its direction of arrival. We need to discriminate against interference. Finally, the larger the antenna the shorter will be the duration of τ of the received pulse and the wider will be the bandwidth $\Delta f \sim 1/\tau$ of the matched filter. If we are using a multi-channel spectrum analyzer this permits us to search a wider band, $B = \Delta f$, for the same number, N, of channels or, if the receiver bandwidth is the limit, to use fewer channels.

Since

$$g \approx \left(\frac{\pi d}{\lambda}\right)^2 = n = \frac{t_s}{\tau} = t_s \Delta f$$

we have

$$d \approx \frac{\lambda}{\pi} \sqrt{t_s \Delta f} \tag{7}$$

Assuming that

t_s = 1 year = 3×10^7 sec
Δf = 6 Hz
T = 10 K
m = 25

we obtain from (5) and (7) the limiting flux levels and antenna diameters shown by the solid lines of figure 3. The sensitivity of 3×10^{-26} W/m^2 at the water hole is quite good: it is only about 43 dB poorer than a full Cyclops. However, the antenna sizes are Cyclopean too. The 1-km diameter at 1.3 GHz would require one hundred 100-m dishes.

There are only two ways to reduce the antenna diameter given by equation (7). One is to reduce t_s, which decreases the sensitivity. The other is to reduce Δf, which requires N to be increased.

If we assume the receiver bandwidth $B = 300$ MHz then, for the case just computed, $N = 300 \times 10^6/6 = 5 \times 10^7$. But we are considering $N = 10^9$, this would permit $\Delta f = 0.3$ Hz and gives the dashed line in figure 3 for antenna sizes.

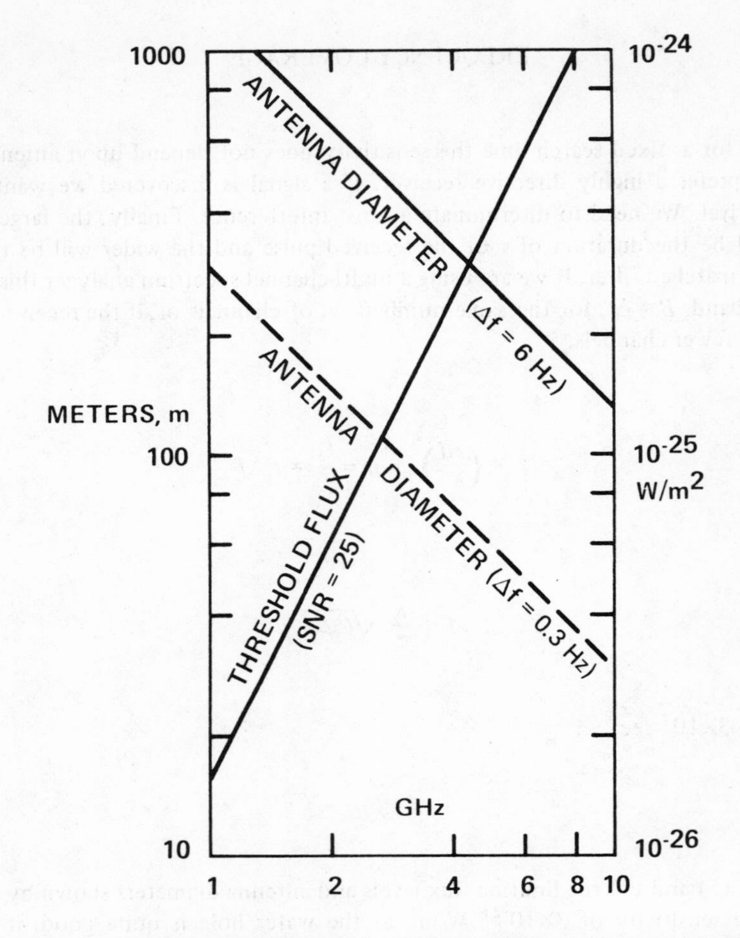

Figure 3.– Antenna size requirements for various flux levels at two channel widths, for a fixed search time of 1 yr.

$A - N$ TRADEOFF

Obviously, as long as the receiver bandwidth is limiting we can trade-off antenna area, A, against the number of channels, N, in the MCSA. Let us find the optimum. At threshold

$$A\phi_o = \Psi\Delta f = \frac{\Psi B}{N}$$

thus

$$AN = \frac{\Psi B}{\phi_o} \equiv \sigma \tag{8}$$

Further, the total cost, C, of the antenna and data processor we take to be

$$C = K_a A + K_c N \tag{9}$$

where K_a and K_c are the dollars per square meter and per channel, respectively. Substituting (8) and (9), and differentiating with respect to A (or N) we find

$$A_{\text{opt}} = \sqrt{\sigma K_c / K_a} \tag{10}$$

$$N_{\text{opt}} = \sqrt{\sigma K_a / K_c} \tag{11}$$

$$C_{\text{min}} = 2K_a A_{\text{opt}} = 2K_c N_{\text{opt}} = 2\sqrt{\sigma K_a K_c} \tag{12}$$

Let us estimate some values for K_a and K_c and see what sort of costs are involved. If we can build a 100-m dish for \$10 million, $K_a = \$1273/\text{m}^2$. If we can build a 10^6 channel MSCA for \$500,000, $K_c = \$0.5$. Assuming that $T = 10$ K and $B = 300$ MHz:

$$A_{\text{opt}} = 4 \times 10^{-9} / \sqrt{\phi_o}\, \text{m}^2 \tag{13}$$

$$N_{\text{opt}} = 10^{-5} / \sqrt{\phi_o} \tag{14}$$

$$C_{\text{min}} = \$10^{-5} / \sqrt{\phi_o} \tag{15}$$

SEARCH TIMES

The preceding designs assume a threshold flux that is independent of λ. Thus, the search times as given by equation (6) will vary as λ^{-2} or ν^2. Rewriting equation (6) in terms of frequency

$$t_s = \frac{4\pi\Psi m}{c^2 \phi_o} \nu^2 \tag{16}$$

The total time required to search from ν_1 to ν_2 where $\nu_2 - \nu_1 \ll B$ is approximately $1/B$ times the integral of equation (16)

$$\Sigma t = \frac{4\pi\Psi m}{3c^2 \phi_o} \frac{\nu_2^3 - \nu_1^3}{B} \tag{17}$$

Taking $m = 25$, and T and B as before, we find from equation (6) that the time to search the water hole is

$$t_{wh} \sim \frac{10^{-18}}{\phi_o} \text{ sec} \tag{18}$$

and to search the microwave window from $\nu_1 = 10^9$ Hz to $\nu_2 = 10^{10}$ Hz

$$t_{\mu w} \sim \frac{5.3 \times 10^{-16}}{\phi_o} \text{ sec} \tag{19}$$

Based upon the above relations (13) through (19) we find the parameters given in table 1 for systems of various capabilities.

TABLE 1.– SYSTEM PARAMETERS

ϕ_o, W/m^2	d_{opt}, m	N_{opt}, M	C_{min}, \$M	t_{wh}, days-yr	$t_{\mu w}$, yr
10^{-23}	70	10	10	1d	1.7
10^{-24}	130	32	32	12d	17
10^{-25}	225	100	100	120d	170
10^{-26}	400	316	316	2.3y	1700
10^{-27}	700	1000	1000	30y	17,000

A few comments are in order:

1. The cost is that of antenna and processor only – it does not include buildings, auxiliary equipment, operating costs, etc.

2. The times assume continuous operation in search mode.

3. Because no Earth-based antenna can cover the entire sky, the possibility of two systems at ±45° latitude should be considered. This would double the costs and halve the times.

4. If arrays are used for the larger size antennas the solid angle of the beam is reduced by the filling factor so the times shown must be divided by the same factor. Two arrays designed to cover the sky from ±45° latitude could have filling factors of 0.9 at 45° elevation and 0.64 at the zenith.

For comparison, the detection sensitivity with 1000 sec observation time per star, vs the number of 100-m antennas and the corresponding costs, assuming $N = 10^9$, are listed in table 2.

TABLE 2.– DETECTION SENSITIVITY AND COST

ϕ_O, W/m^2	Number of antennas	Cost, $M
10^{-27}	2	70
10^{-28}	20	250
10^{-29}	200	2,050
10^{-30}	2000	20,050

The time required to search the water hole for all F, G, and K stars within

100 light years = ~1 month
1000 light years = ~50 years

CONCLUSIONS

The time required to search the entire sky for CW signals is inversely proportional to the minimum detectable flux level and, for large relative bandwidth, increases as the cube of the highest frequency. By replicating optimum systems, dollars can be traded for time. If enough systems are built to keep the search time constant the cost varies as the 3/2 power of the sensitivity.

A full-sky search of the water hole could be made to a flux level of 10^{-24} W/m^2 in only 12 days with an optimum system. To search the entire 1 to 10 GHz region to the same sensitivity would require 17 years, half of which would be spent searching between 8 and 10 GHz. Evidently if the entire microwave window is to be searched, the sensitivity should be reduced as the operating frequency increases. If the minimum detectable flux level is made proportional to ν^2 the search time is proportional to the frequency range covered rather than the cube. This would permit covering the 1 to 10 GHz region at from 5×10^{-25} to 5×10^{-23} W/m^2, respectively, in about 1 year. This assumption of Φ_O proportional to ν^2 is consistent with the assumption that the signal we will detect is being beamed at us, for then $\Phi = P_T A_T / R^2 \lambda^2$. Hence the full sky search covers the cases of beamed or very strong omnidirectional signals originating at great distances: the "giant" or "supergiant" transmissions of very advanced cultures.

APPENDIX A

Assume that the amplitude of the illumination of a circular antenna of radius b is

$$u_\nu(r) = u_o\left(1 - \frac{r^2}{b^2}\right)^{\nu-1} \tag{A1}$$

These are the so-called Sonine functions. The effective area is

$$A_{\text{eff}} = \frac{(\int u\ dS)^2}{\int u^2\ dS} = \frac{2\nu - 1}{\nu^2}\ \pi b^2 \tag{A2}$$

and the on axis gain is

$$g_\nu = \frac{4\pi A_{\text{eff}}}{\lambda^2} = \frac{2\nu - 1}{\nu^2}\left(\frac{2\pi b}{\lambda}\right)^2$$

The decreasing effective area and gain with increasing ν are the result of illumination being confined more and more to the central region. We can compensate for this by letting

$$b = \frac{\nu a}{\sqrt{2\nu - 1}} \tag{A3}$$

where a is independent of ν. The clear aperture case corresponds to $\nu = 1$. The scaling given by equation (A3) gives every case the same effective area and gain as a clear aperture of radius a. If we now let $\nu \to \infty$, equation (A1) approaches

$$u(r) = u_o e^{-2r^2/a^2} \tag{A4}$$

The far field amplitude patterns are the Hankel transforms of (A1) and (A4) and, for finite ν, are

$$f_\nu(\theta) = \Lambda_\nu(k\theta) \tag{A5}$$

where $k = 2\pi b/\lambda$ and $\Lambda_\nu(z) = \nu!\ 2^\nu J_\nu(z)$. For the Gaussian case $\nu = \infty$

$$f(\theta) = e^{-g\theta^2/8} \tag{A6}$$

where $g = (2\pi a/\lambda)^2$ is the on-axis gain. The half-power beamwidth, β, is twice the value of θ that makes $\Lambda_\nu^2 = 1/2$ or $e^{-g\theta^2/4} = 1/2$. We find

ν	$\beta(\pi a/\lambda)$
1	3.23268
2	3.45443
3	3.44851
4	3.43173
∞	3.33022

If we tessellate the sky with hexagons of width B between opposite sides, each will have a solid angle $\Omega = (\sqrt{3}/2)\beta^2$ so the required number is $n = 4\pi/\Omega$. For the cases listed

$$1.22\, g \leqslant n \leqslant 1.39\, g$$

If we scan the sky in strips of width β, the effective dwell angle per direction is

$$\Delta\theta = \int_{-\infty}^{\infty} f^2(\theta)\, d\theta \tag{A7}$$

For the Sonine cases this gives

$$\Delta\theta_\nu = \frac{2^{3\nu+1}\, (\nu!)^2\, (2\nu - 1)!}{\pi[1\cdot3\cdot5\cdot\cdot\cdot(2\nu-1)]^2\, [1\cdot3\cdot5\cdot\cdot\cdot(4\nu-1)]}\, \frac{\lambda}{2\pi a} \tag{A8}$$

For the Gaussian case

$$\Delta\theta = 2\sqrt{\pi}\left(\frac{\lambda}{2\pi a}\right) \tag{A9}$$

ν	$\Delta\theta\,(2\pi a/\lambda)$
1	3.3953
2	3.5845
3	3.5899
4	3.5848
∞	3.5449

The effective number of directions in the scanning mode is

$$n = \frac{4\pi}{\beta\Delta\theta} \tag{A10}$$

Thus for various ν we find

ν	n
1	1.145 g
2	1.015 g
3	1.015 g
4	1.021 g
∞	1.064 g

and the assertion that $n \approx g$ seems justified.

REFERENCE

1. Oliver, Bernard M.; and Billingham, John: Project Cyclops, A Design Study of a System for Detecting Extraterrestrial Intelligent Life. NASA CR 114445, 1972.

Prepared by: Bernard M. Oliver
Vice-President for Research
and Development
Hewlett-Packard Corporation

Two parabolic reflector antennas, forming the research and development site of NASA's worldwide Deep Space Network, stand out vividly against the primitive beauty of Southern California's Mojave Desert. The bowl-shaped location provides natural protection from man-made radio interference. Both antennas are equipped with Cassegrain feed systems and are steerable from horizon to horizon. The dish diameters are 26 m (85 ft) and 9 m (30 ft). The larger uses a low noise maser preamplifier to achieve a very high sensitivity for an antenna of this size. Antennas such as these can be used for sky and frequency SETI surveys. Depending on frequency, such antennas can perform all-sky surveys over plausible microwave regions with several orders of magnitude better sensitivity than has been generally achieved. The Deep Space Network is managed and technically directed for NASA by the Jet Propulsion Laboratory of the California Institute of Technology at Pasadena, California.

4. STELLAR CENSUS

Operation of a target-search strategy requires a target list. Even if we knew precisely what types of stars were accompanied by planetary systems believed likely to provide favorable life sites, we would not know where to look for the bulk of these stars in the neighborhood of the Sun, that is, within a range of 10^3 light years (ly). Present catalogs out to 10^3 ly are probably incomplete for F, G, and K dwarf stars by a factor of 10^3. For early M dwarf stars, the situation is worse. As long as we favor planets orbiting stable stars as the probable sites for intelligent life, so long will we be in essential need of a Whole Sky Catalog, or stellar census of stars down to the 14th or 15th magnitudes. Even then the catalog would be seriously incomplete for M dwarf stars.

A suitable stellar census should have the following properties:

1. A high degree of completeness out to at least 1000 ly. Because planets orbiting early M dwarf stars (M0–M4) are generally not excluded as possible life sites, the census should extend down to at least the 15th apparent visual magnitude. As a consequence, 25×10^6 or more stars must be identified.

2. In the main, each star should have its MK classification established to a satisfactory degree. Where feasible, more refined classification is highly desirable. In particular, it is desirable to provide estimates of stellar age.

3. Stellar position and radial distances of each star should be determined to the greatest precision consistent with completing the first edition of the census in about a decade.

4. The census should be cross referenced to previously existing catalogs.

5. Where known, additional information such as parallax, proper motions, duplicity, variability, etc., should be included. In fact, the initial census should, to the degree possible, be considered both as a summation of current stellar information and as first epoch observations for a massive improvement in our knowledge of the stars in our neighborhood. In the process, of course, similar information on an enormous number of stars at much greater distances will also be obtained.

It is the contention here, though yet unsubstantiated by thorough study, that such a census can be achieved at modest cost in about a decade following funding. The approach we suggest has the following major ingredients:

1. Digitize about 400 standard MK spectra. If necessary, take additional spectra at suitable dispersion.

2. Carry out a multivariate (or multifactor) analysis seeking the optimum achievable color systems using precision photographic photometry. Assess the error contributions as well as the capabilities of the system.

3. Design a completely automated photographic photometric system, telescope to tape data bank. The design should avoid manual operations other than transporting developed plates or, preferably, sheet film, in proper containers, from the observatory to the automatic plate reading machines. Automatic standardization and calibration procedures should be provided. Separate small telescopes should be used to obtain extinction information. Moderate field telescopes of the order of 60 in. should be adequate. Above all, the design of the telescope, its controls, the auxiliaries, the photometric apparatus, and the domes should be considered to be a single integrated system design problem.

4. Install identical systems at optimum sites, one in the Northern Hemisphere and one in the Southern Hemisphere.

5. Let all photographic data be measured by computer controlled machines. The same computer can correlate, classify, and store the data, and assess the results in real time. That is, completion of the catalog should about coincide with completion of the necessary observations.

Such a stellar census would have much value to general astronomy. Discussions with some photometric and spectroscopic specialists have strongly supported the belief that such a system is feasible. In any event, laboratory trials at an early stage of the design study will surely clarify the situation.

Should photographic methods prove unsatisfactory, it will be necessary to develop an appropriate photoelectric sensor system. This is clearly possible, and should be explored as a technology requiring a special development effort.

Prepared by: Charles L. Seeger
SETI Program Office
Ames Research Center

5. SUMMARY OF POSSIBLE USES OF AN INTERSTELLAR SEARCH SYSTEM FOR RADIO ASTRONOMY

INTRODUCTION

Radio astronomical investigations of great scientific interest can be carried out with the wide range of SETI antenna systems presently under discussion. This range includes both the SETI programs planned for the near future with existing antennas, and larger ground-based or space-borne antenna systems that might be built in the future.

The effect of SETI technology on radio astronomy can be broadly broken down into two classes. In the first class we have the extension or improvement of existing microwave technology:

1. *Receiver design:* A class of receivers to be developed for SETI is characterized by near optimum noise figures ($\sim$10 K at a room temperature waveguide flange), broad instantaneous bandwidth ($\gtrsim$300 MHz), and octave bandwidth tuning ranges. This technology will probably be rapidly adopted by radio observatories so that the possession of such receivers will not make SETI systems unique, but would be a SETI spin-off.

2. *Collecting area:* Eventually, a SETI receiving system may vastly surpass radio astronomy facilities, existing or projected, in collecting area. (Compare the VLA at twenty-seven 25-m antennas to even two 100-m antennas.)

In the second class of SETI impact we have the development of a new generation of signal processing facilities:

3. *Data Processing Hardware:* On the basis of SETI requirements, it is possible to predict the general properties of such hardware. To make a microwave search tractable, it will be necessary to utilize fully the entire receiver bandwidth ($\gtrsim$300 MHz) while retaining high spectral resolution: a processor of $\geqslant 10^6$ channels. An integrating spectrometer with such characteristics is an impressive scaling-up of present radio astronomy technology, but a significant development is called for when we admit that we have no a priori knowledge of the nature of SETI signals. Then, we require a fully flexible data processing system that can measure all properties of a signal (e.g., frequency distribution, time structure, polarization) continuously in real time. The ability to fully characterize radio signals offers hope for recognizing and rejecting various kinds of interference. If the processor is eventually to be used with an antenna array, the ability to operate several array subsets or several array beams simultaneously would be very useful.

4. *Data Acquisition Management:* The management of a 10^6 channel signal processor and the extraction and sorting out of various kinds of scientific data in real time represents another breakthrough area. The following kinds of data are some that need to be managed, preferably simultaneously:

- Time averaged spectrum (10^6 channels; recognize and extract spectral lines for astronomy; Doppler compensate for space motion of observatory)

- Dynamic spectrum (10^6 channel with high time resolution; search for Doppler patterns perhaps characteristic of ET transmitters; recognize interfering signal patterns; e.g., solar bursts in sidelobes, satellite transmitters, etc.)

- Polarization (four Stokes parameters in 10^6 channels needed for signal recognition; i.e., a weak, narrow, unpolarized signal would probably be an unknown spectral line rather than an ET signal; an ET signal might be polarization modulated)

- Dispersion removal (an ET signal might consist of broadband pulses that would be dispersed in the interstellar medium)

Thus, this second class of SETI impact would represent a major new way of handling data. It would permit astronomers to engage in survey projects of a scope that has only been attempted a few times in the past, and then only with a large dedication of scientific manpower. In this capacity, a SETI system would likely be unique for a considerable period of time.

As SETI activities widen in scope and increase in sensitivity, the utility of SETI facilities for radio astronomical investigations will surely increase. In particular, aspects (2–4) above present enormous potential for improvement. It is also apparent that immediate SETI efforts utilizing currently achievable advances along the lines of (1) and (3) will yield new results of significant radio astronomical interest. This complementary document discusses specific scientific benefits that would arise from SETI efforts. This treatment is by no means exhaustive. For instance, serendipity is a vital factor attendant to any major leap in instrumentation. It is, however, impossible to discuss benefits that derive from new and unexpected discoveries. As in the case of the 200-in. Hale Observatories telescope, there will surely be many that derive merely from each significant increase in collecting area. In addition, extensive sky and frequency coverage with high frequency resolution (several Hz, or ~0.001 km sec^{-1} at 1.5 GHz), wide instantaneous bandwidth (~300 MHz), and possible sensitivity to pulsed signals will surely result in new discoveries of scientific importance.

Remaining within the domain of foreseeable scientific benefit, we present here likely applications of several near- and far-term SETI systems to radio astronomy. Three different scales of system complexity are represented: an optimally equipped single 26-m antenna, the equivalent of the full Cyclops array of 1026 antennas each of 100-m diameter, and an intermediate case.

POTENTIAL SCIENTIFIC APPLICATIONS OF A 26-METER SETI SYSTEM[1]

Astronomical investigations of individual radio sources have achieved higher sensitivity levels than could be obtained with a 26-m antenna in a survey program. As a survey instrument, however, a 26-m SETI facility with an optimum front end compares favorably in sensitivity with surveys that have been done, but with the added advantages of higher spectral resolution, greatly expanded frequency coverage, and complete coverage of the visible sky (see Section I-2).

In considering potential programs, we will assume a minimal sensitivity system consisting of a 15 K system on a 26-m antenna. We will assume a dual-polarization receiver, although the same sensitivity can be achieved with a single-polarization system operating twice as long. In comparing various observations, we will compute the minimum detectable flux density from

$$S = 1.757 \times 10^4 \frac{K_D K_R}{\eta_R} T_s D^{-2} (B\tau)^{-1/2}$$

where S = flux density, Jy; K_D = detection limit factor ≈ 5; K_R = receiver mode factor, $\sqrt{2}$ for receiver switching, $\pi/2$ for autocorrelation spectrometer, $\sqrt{2}$ if τ includes both *on* and *off* measurement; T_s = system temperature, K; D = antenna diameter, m; B = bandwidth, Hz; τ = integration time, sec; and η_R = antenna aperture efficiency.

For studies of extended objects, such as the larger interstellar clouds, the antenna resolving power may not be an important factor. For each sky position, we have a minimum detectable brightness temperature of

$$T_B = \frac{K_D K_R}{\eta_B} T_s (B\tau)^{-1/2}$$

where η_B is the beam efficiency. A higher sensitivity can be achieved by averaging adjacent sky positions so that the effective beamwidth is larger.

Radio Source Surveys from SETI: Number-Flux Density Relationship and Spectra of Sources

A natural consequence of the SETI program will be a number of very sensitive radio source surveys over the frequency range 1.4–23 GHz covering all the visible sky. The sensitivity that will be achieved in the constant beamwidth surveys extends beyond the confusion limits for nearly all frequency intervals which would be observed with the 26-m telescope. Thus it will be possible to

[1] This section includes contributions from S. Gulkis, M. Janssen, T. Kuiper, and E. Olsen of Jet Propulsion Laboratory, Pasadena, Calif.

generate radio source surveys over 6 sr of sky which are complete to the 0.3 Jy level[2] (or less) for any frequency desired in the range quoted above. For comparison, the NRAO "deep" survey at 5 GHz, carried out using the 43-m telescope, is complete to the 0.1 Jy level over only 6×10^{-3} sr of sky (ref. 1).

In particular, all-sky surveys at high frequencies have not been carried out in the past. The larger sample of sources that would be available through the SETI program would give greater statistical accuracy to the source counts and to the distribution of the sources within the different optical and radio classes that have so far been found. Surveys at very high frequencies will aid the study of the population of very young sources, and may even uncover new classes of sources. A uniformity of beam size will be especially useful in the intercomparison of surveys at different frequencies.

Radio Recombination Lines

When ionized hydrogen gas recombines, radio spectral lines are emitted when the atom passes through states of high excitation (large quantum number n). For example, between 1.4 and 1.7 GHz, transitions between quantum levels 157 to 166 are observed. The phenomenon is most prominent in hot, ionized gas near early-type stars. A survey of 43 sources was conducted by Dieter (ref. 2) at 1.65 GHz, using the Hat Creek 26-m antenna. The sensitivity limit of this search ($K_R = \sqrt{2}$, $T_s = 140$ K, $B = 10^4$ Hz (1.8 km sec^{-1}), $2^h < \tau < 20^h$) varied between $3.0/\eta_R$ and $0.96/\eta_R$ Jy, depending on the source observed. With an optimized 26-m SETI system, the same sensitivity will be reached in 1.5 to 15 min. However, if one considers averaging the results for all the recombination lines in this 300 MHz band, sensitivity is increased by a factor of 3.

If the simplifying assumption is made that the ionized gas is populated according to a Boltzmann distribution, combining a measurement of free-free continuum emission with a recombination line intensity allows us to deduce the electron temperature and density. This has already been done for 120 sources in the northern sky by Reifenstein et al. (ref. 3) using a recombination line at 5 GHz. A SETI-related all-sky survey using a 26-m antenna would not add significantly to this. However, it is known that in some sources at least, the population of energy levels does not follow a Boltzmann distribution. This effect, which is due to strong radiation fields and low collision rates, can be determined by measuring recombination lines at widely separated frequencies. For this reason, recombination line data from a SETI survey covering many frequencies would be useful.

As the electron temperature of an ionized region decreases, the recombination line intensity relative to the radio continuum increases. Thus, narrow recombination lines might be observed in the absence of radio continuum emission, although calculations for plausible situations suggest that very high sensitivity would be required (integration measured in days). If the electron density is also low, narrow lines could arise by stimulated emission. Weak narrow lines have been observed

---

[2] One Jansky = 10^{-26} W/m^2-Hz

in a number of instances (ref. 4) although whether the former or latter mechanism is responsible is not clear. Discovery of a strong, narrow line in an unexpected direction would be of high scientific interest. Ruling out the presence of such lines would be useful.

Neutral Hydrogen

Galactic Hydrogen– Within the general velocity limits of ±100 km sec^{-1} and at a frequency resolution of about 10 kHz, neutral hydrogen has been very extensively mapped. Without repeating current survey work, there is interest in searching for narrow features, most likely to occur in absorption. Knapp (refs. 5 and 6) has observed selected dust clouds with good velocity resolution (0.08 – 0.34 km sec^{-1}). Somewhat less than half of the clouds showed such features. This suggests that an all-sky survey with high frequency resolution will be useful as a technique for mapping cold dust clouds. The shapes of these absorption features can be analyzed for internal cloud motions, and their intensity can be compared with visual absorption in further studies of the gas/dust ratio in local cloud kinematics and cloud structures.

The study of high-velocity clouds, with bearing on theories of galactic structure, would benefit greatly from an all-sky 26-m survey. These clouds are widely distributed, and have velocities as large (in one case) as 400 km sec^{-1}, or a frequency shift of ~2 MHz. Most surveys have been more restricted in velocity space, and have covered only limited regions of the sky. In addition, many known high-velocity-clouds have narrow frequency half-widths, some being unresolved by existing surveys. Thus an all-sky, broad bandwidth, high spectral resolution survey would almost certainly turn up new and interesting results on these interesting objects.

Globular Clusters– Neutral hydrogen observations of globular clusters have been conducted, both to examine the content of the clusters themselves (ref. 7) and to study the nature of the intervening clouds (ref. 8). With the sensitivity of a SETI search, these data would be made complete.

Extragalactic Hydrogen– Recently, Fisher and Tully (ref. 9) made neutral hydrogen observations of 241 extragalactic systems in the David Dunlap Observatory (DDO) catalog. These are dwarf-like systems of low surface-brightness found on the Palomar Sky Survey. The observations were primarily with the NRAO 91-m telescope, and achieved a sensitivity (T_S = 50 K, τ = 240 sec, $K_R = \pi/2\sqrt{2}$, $0.45 < \eta_R < 0.55$) of ~0.55 Jy for a bandwidth of 100 km sec^{-1} (4.73×10^5 Hz). Altogether, 179 systems were detected. This led to re-examination of the Palomar Sky Survey to obtain a candidate list of more than one thousand systems, of which more than two thirds have been detected in neutral hydrogen. This suggests further productive searches can be done.

A highly optimized 26-m system compares favorably with the NRAO 91-m antenna. In favor of the 26-m system are a drastically lower system temperature, a sensitivity increase of $\sqrt{2}$ from the use of two polarizations, a sensitivity increase of $\pi/2$ from using a spectrometer which does not clip the signal (as the NRAO autocorrelator does), and a somewhat higher aperture efficiency. Taking these factors into account, we can achieve the same sensitivity as Fisher and Tully did by integrating 8 min per point rather than 4 min per point. Another way to look at this is that if we scan the sky at a sidereal rate ($130^s \sec \delta$), we would have roughly half the sensitivity of Fisher and Tully, and detect 60 percent of their initial list. The usefulness of the survey drops off very rapidly with diminishing sensitivity. With one polarization, 38 percent of their sources would be detectable; if the effective time per position is reduced to 65^s, 21 percent of their sources would be detected.

There are a number of interesting possibilities for an all-sky survey. We would detect systems that are not optically visible, either because their surface brightness is intrinsically low or because they are obscured by material in our own Galaxy and, of course, we would measure their redshift directly from the radio line. Ultimately this would lead to a much better understanding of the mass function for galaxies, and the correlation of neutral hydrogen mass with system type.

Another very intriguing possibility is the detection of small condensations of neutral hydrogen within our own local group. In a search at the sidereal rate, our detection threshold would be $3 \times 10^6 D^2 M_\odot$, where D is the distance of the condensation in megaparsecs. (The Magellanic Clouds are at ~0.06 Mpc distance; M31 and M33 at 0.7 Mpc.) The detection of such material would, of course, be very exciting. Statistical studies of the redshifts would define the rest standard of the local group, and a measure of the "temperature" of the early universe from the dispersion in the local velocities. Even a nondetection would be useful in determining whether the local group is bound and whether there is a lower limit to the mass function of galactic systems.

This potentially exciting program puts severe constraints on the system. Absolutely every contribution to system noise would need to be minimized, and advantage taken of every opportunity to improve our statistics (e.g., dual polarization). The fastest observations would be at a sidereal rate (~200 days for the whole sky) and preferably, we would take longer. The minimum useful instantaneous frequency range to be examined (-3000 km sec^{-1} $< v <$ 1000 km sec^{-1}) is 20 MHz, while 30 or 40 MHz would be desirable.

Hydroxyl Radical (OH)

Emission from the hydroxyl radical has been observed (primarily at the frequencies of 1612, 1665, and 1720 MHz) from a wide variety of sources. It is found extensively throughout dust clouds and appears in masers in such a variety of sources as HII regions, planetary nebulae, late spectral type stars, Wolf-Rayet stars, infrared sources, and supernova remnants. Within this wide range of objects, the OH excitation shows considerable variation, and Turner (ref. 10) has provided a classification scheme that is able to correlate the nature of the source with its OH spectral character for a large number of sources. Currently, Turner is reducing a survey of OH sources that covers about one-third of the galactic plane. It is essentially complete ($337^\circ < \ell < 270^\circ$, $|b| < 1^\circ$) for a 1° strip, about 30 percent complete for the adjacent 1° strips, and has random coverage for selected objects in the rest of the sky. Altogether, about 2000 points were observed over the velocity range ±120 km sec^{-1} relative to the galactic rotation velocity. However, larger velocities are known to occur so that the greater velocity coverage in a SETI survey would be of scientific interest.

With a 26-m SETI system we would have roughly comparable sensitivity. An extension of the Turner survey to the whole sky and over an extended velocity range would be very useful. The data would be applicable to studies of the nature of OH excitation, studies of galactic kinematics, cloud structure studies, and in searches for star formation sites.

OH emission is generally polarized. A survey would therefore be most effective in a dual polarization mode, employing a data processing system that would yield all four Stokes parameters. The Zeeman effect has been proposed to account for the polarization of OH line emission. A 26-m SETI all-sky survey would observe a large sample of OH sources for apparent Zeeman patterns, yielding a useful probe of the local magnetic field strength in the Galaxy. The direct measurement of the magnetic field strength in regions of star formation will have an important bearing upon our understanding of the processes involved.

Methyladine Radical (CH)

The presence of CH and its ion, CH^+, in the interstellar medium has been known since their detection at optical wavelengths in 1937 at Mount Wilson. While the radio study of this molecule is still in its infancy, it promises to be an important tool for studying the interstellar medium, particularly because of the apparent sensitivity of its abundance to local density (ref. 11).

A 26-m SETI survey with the assumed system parameters (T_s = 15 K, K_R = 2) would achieve a sensitivity comparable to the current work (rms ~ 0.005 K, $B = 10^4$ Hz) in approximately 1 hr of integration per sky position. However, useful information would still be obtained with integration periods of the order of several minutes, particularly in terms of locating the regions of most intense emission.

Formaldehyde

Formaldehyde is one of the most useful probes of interstellar clouds. Primarily by its transition at 4.83 GHz, it has been used to define the extent and distribution of interstellar clouds. The distribution of formaldehyde correlates well with the distribution of dust (refs. 12 and 13). Thus, it has been possible to study the kinematics of dust clouds in the solar neighborhood (ref. 14). Since the 4.83 GHz transition has sufficiently separated hyperfine components, it has been possible to use the hyperfine ratios to estimate the optical depth of the clouds. Because formaldehyde occurs in denser regions than neutral hydrogen, it is generally useful for the study of specific regions such as the galactic center and regions of star formation, and even for discovering major cloud complexes not optically visible (ref. 15).

Even the smaller radio astronomy facilities (e.g., Hat Creek 26-m) have an antenna beamwidth at this frequency ($\lesssim$10 arcmin) which makes an all-sky survey impractical. However, an all-sky survey of medium resolution (30 arcmin) obtained by averaging adjacent records would be very useful in providing an overall picture of the distribution of dust clouds in the Galaxy, and may, in fact, be more effective than current HI surveys in locating the spiral arms of the Galaxy (ref. 16).

Ammonia

Interstellar ammonia has not been extensively studied, largely because the transitions are quite weak, a few tenths of kelvins. The most complete study is that of Morris et al. (ref. 17), which indicated that ammonia is fairly widespread in the interstellar medium. This molecule arises in regions having densities $n_{H_2} > 10^4$ cm^{-3}. It also has the attractive feature of having a large number of transitions within a narrow frequency range (23.6 to 25.1 GHz), several of which are between metastable rotational levels. Thus, it would appear that ammonia is potentially as useful as CO as a probe of the temperatures of interstellar clouds. With an optimally equipped 26-m antenna, an all-sky survey at the ammonia frequency would readily detect many new sources.

Water

Water shares with hydroxyl the importance of being an indicator of some unusual processes occurring in the extended envelopes of certain kinds of stars. Unlike hydroxyl, however, it is seen only in maser action. Because even small radio telescopes have quite small beamwidths at this frequency (22.235 GHz), the total sky coverage of all H_2O observations is quite small, and it is quite likely that the present set of known H_2O sources is quite biased.

It appears that a SETI survey would make two important contributions. Because H_2O maser lines are quite strong, a broad beam survey could locate intense H_2O sources that have not been included in the objects examined to date. Also, a sensitive targeted survey of selected stars would broaden the classes of stars examined and thus lead to a more exact understanding of the kinds of stars that are associated with H_2O emission.

Radio Source Polarization Studies

Extragalactic Radio Sources– Here the main strength of the SETI program would be its ability to investigate all four Stokes parameters over a very wide range of frequencies (e.g., 1 to 25 GHz) in extremely fine frequency steps. In particular, all previous studies have been carried out using broadband, double-sideband receivers. The IF bandpasses of these receivers have rarely been as small as 20 MHz, and never smaller than 10 MHz. Hence, the fine-scale polarization structure of microwave spectra is completely unknown. It is quite likely that microstructure is present, arising from differential Faraday rotation and depolarization among several localized domains in these sources. At this time the study of magnetic field structure of extragalactic radio sources can be carried out for a very small number of extended sources at a few frequencies using polarization interferometry. SETI would allow statistical studies of inhomogeneities in the magnetic field structure of many unresolved sources.

Galactic Radio Sources and Galactic Magnetic Field– Discrete continuum radio sources in our own galaxy which emit polarized radiation, such as supernova remnants, would be investigated in the same manner as the extragalactic sources.

The continuum background nonthermal emission originating in the galactic plane has been investigated at a number of wavelengths, all greater than 20 cm. At 21 cm the nonthermal emission in the galactic plane amounts to a brightness temperature of at least 1 K. If we assume a HPBW of 1° and a bandpass of 300 MHz, the available nonthermal flux is approximately 4×10^{-17} W/m². If the polarized flux is only 1 percent of the total, an optimized 26-m antenna would achieve a significant signal-to-noise ratio in each beamwidth solid angle while scanning at the sidereal rate. An all-sky survey of the SETI variety will automatically generate maps of rotation measure and intrinsic position angle for galactic nonthermal emission; these in turn would delineate structure of the longitudinal and transverse components of the galactic magnetic field and would extend spectral coverage to shorter wavelengths as well.

Intergalactic Magnetic Field in the Local Group– The all-sky survey, if carried out to high sensitivity, may be able to detect the Faraday effect due to the intergalactic magnetic field. This would be an extremely difficult measurement.

Pulsars

The time-averaged flux density of pulsars is generally quite weak: at 1.5 GHz it is ≤0.1 Jy. This may be compared with a minimum detectable flux for a 26-m SETI program ($T_S = 15$ K, $\tau = 200^s$, $K_R = 1$, $\eta_R = 0.6$) of 45 $B^{-1/2}$ Jy. So, pulsars start to become detectable at a bandwidth of 200 kHz, and at 20 MHz, we have good signal-to-noise for the stronger ones. The wider bandwidths can only be achieved, however, if the dispersion (differential time delay) is removed. Thus, the observation of known pulsars at estimated levels of sensitivity would not augment current knowledge significantly.

There is the possibility of detecting new pulsars but two difficulties are encountered. The first is that the flux of pulsars diminishes typically with the second to third power of the frequency. Detecting pulsars at 1 GHz would be at least twice as easy as at higher frequencies; a limiting sensitivity is reached approximately at that frequency where the galactic background dominates the receiver noise. However, previous surveys (though probably more sensitive) may be incomplete in sky coverage or dispersion measure range, so that the value of a 1.4 GHz survey bears further investigation.

The second problem associated with a pulsar search is that we must search two new parameter regions: pulse repetition rate and dispersion. However, dispersion removal techniques exist and would be readily implemented (ref. 18). The removal of dispersion effects is relevant not only to pulsars, of course, but to all pulsed signals, whether of natural or intelligent origin, and would be a valuable part of any projected data processing system.

POTENTIAL SCIENTIFIC APPLICATIONS OF LARGE-SCALE SETI SYSTEMS

Approach

At the Third Science Workshop on Interstellar Communication, a group of radio astronomers[3] evaluated the Cyclops array as to its effectiveness in answering current astronomical questions.[4] This complementary document presents those evaluations in a concise format, and in addition, attempts to evaluate the effectiveness of more modest SETI systems, of a scale intermediate between an equivalent to a full-scale Cyclops and a single antenna. This evaluation is tabulated in the charts presented at the end of this section.

The radio astronomical frontier, as envisioned by the speakers at the Third Science Workshop, was broken into topical areas and the needs of studies in these areas were evaluated in terms of instrumentation. That is, four dimensions of the instrument – (1) sensitivity (total area), (2) spatial resolution (linear extent), (3) frequency range (minimum and maximum useful frequencies), (4) frequency resolution (i.e., how narrow) – have been considered as to whether they may have threshold or limiting values for useful work to be done in the topical areas. The method was simply to scale values of parameters given by the Workshop speakers (flux density, integration time, distance, etc.) to values characteristic of the use of more modest instruments. These parameters are related to each other by the inverse-square radiation law and the classical radio astronomy sensitivity equation,

$$S = KT_S/A\sqrt{tB}$$

where S is the minimum flux density detectable by a given system, T_S the system temperature, A the effective collecting area, t the integration time, B the system bandwidth (assumed here to be less than the "signal" bandwidth), and K a constant. Where no numbers were given by the Workshop speakers, no scaled values are presented here. Here we will deal primarily with variations in (1) collecting area, and to some extent (2) spatial resolution.

Not all of the evaluations presented at the Third Science Workshop are "topical areas" in the sense of scientific "questions." For instance, use of a SETI system as a VLBI terminal or for polarization studies are techniques and not current astronomical questions, and thus evaluations from these standpoints are distributed throughout the analysis by topical area.

---

[3] A list of those who spoke on the use of a SETI system for radio astronomy is given in Section III-15.

[4] Parameters of the full Cyclops array: 3.2 km clear aperture diameter, 10 km array diameter, 1026 antennas at 100 m diameter. At 1 GHz, single antenna efficiency = 60 percent, system temperature = 20 K, array beamwidth = 7″, single antenna beamwidth = 13′ (Bracewell, Third Science Workshop). Note that the choice of these specifications for a "full" Cyclops array is essentially arbitrary, and is made for convenience. A Cyclops system may in fact consist of any number of antennas. In fact, it is the number required to detect signals of extraterrestrial intelligent origin.

Before discussing the findings of this summary, we will briefly elaborate on "confusion," one of the constraints noted in the charts. At each frequency, and for each instrument, there is a confusion-limit to the minimum detectable flux density of a postulated discrete object. It is that flux density at which one finds at least one extraneous weak object in the main beam of the instrument (or one strong object in a sidelobe) and is thus a function of spatial resolution. As there are more objects per steradian at lower fluxes, a smaller beam must accompany studies of weaker objects. A thorough treatment of confusion is beyond the scope of this brief discussion.[5] However, using a likely frequency scaling relationship and a value of 10^{-7} Jy as the confusion limit at 3 GHz (quoted by Kellerman for the Cyclops array) one may easily determine that confusion will be the limiting factor in many of the studies mentioned in the charts. Thus the resolution of the "nominal" SETI system will need to be increased for successful attainment of those quoted detections for which confusion is given as the limiting factor. This is easily done by the addition of "outrigger" array elements spaced at several array diameters from the main array. In the following discussion, proper use of outrigger antennas is presupposed.

Discussion and Summary of Findings

In a general fashion, one may reach two interesting conclusions from even such a preliminary analysis as is given in the charts at the end of this paper.

First, it appears that the combination of sensitivity (surface area) and high resolution will make a large-scale system especially valuable to advances in our understanding of galactic phenomena, and in particular, to the following.

1. *The interstellar medium*: The "Hierarchy" of clouds (densities, temperatures, molecular abundances, kinematics, the role of magnetic fields in cloud collapse), especially if the frequency range to 25–30 GHz is obtainable.

2. *Stellar system formation and evolution*: Protostellar systems, T Tauri variables, stellar winds in general, flare stars, evolved stars, peculiar broadband time variable stars, especially if a time resolution capability or frequency range to 30 GHz is provided. Knowledge of the structure and perhaps composition of the atmospheres and surfaces of the planets of our own solar system would be greatly enhanced.

3. *Galactic structure*: Small-scale distribution of hydrogen in our Galaxy and detailed mapping of nearby galaxies to an extent that would allow refinement of theories of galactic structure.

4. *Pulsars*: Both in the sense of enhanced knowledge of the pulsars themselves, and in the sense of the very useful tool for mapping the small-scale properties of the interstellar medium that they would then provide. The dispersion removal problem is more tractable by a factor of

---

[5] See, however, reference 20 for a thorough discussion that includes the likely frequency dependence.

$(f_H/f_L)^2$ at higher ($f_H = 1$–10 GHz) frequencies than at the frequencies ($f_L \sim 100$ MHz) at which pulsars are strongest and are currently studied.

In addition:

5. *Studies of extragalactic objects of low brightness* (bridges between possibly associated objects, etc.) will also be greatly enhanced, especially if outrigger or similar (VLBI) resolution enhancing capability is provided, as will studies of "clumping" within galactic clouds.

6. *Cosmological studies* would be furthered by at least one important experiment; the accurate determination of the redshift-magnitude relation for normal galaxies out to $Z = 1$, with the possibility of other important observations (see chart).

However:

Studies of small, bright objects (QSO's, radio galaxies, etc.) are perhaps better suited to (and perhaps influence the design of) VLA-type instruments. Such instruments and objects are very nicely complemented in their capabilities and topics of study by any foreseeable SETI system.

Second, it appears that intermediate-scale SETI systems (10–30 percent of the magnitude of the full Cyclops array) are able to address practically all of the questions comprising our radio astronomical frontier as effectively as the full Cyclops. Longer integration times are of course required of more modest systems. The current questions listed by the Third Workshop speakers would, in most cases, be answered by such a "partial Cyclops" system within the useful lifetime of the instrument (decades).

These same questions would be answered by a system of scale comparable to the full Cyclops in less than a year. It is surely unimaginable that an instrument of this sensitivity would remain idle after answering all of our current questions; it has been the history of science that each answer becomes in its turn a new question.

Introduction to Chart

This chart represents, primarily, a summary and organization of the statements made by speakers at the Third Science Workshop on the uses of a proposed SETI system (full Cyclops) to radio astronomy. Scaling is given in the last two columns.

CHART I.– CAPABILITIES OF LARGE SETI SYSTEMS FOR CURRENT RADIO ASTRONOMICAL RESEARCH

Topic	Subhead	Instrumental capabilities or observations needed	FC capabilities	Possible		PC	
				Thresholds	Limits	Dimensions	Capabilities
Cosmology	Isotropy and homogeneity of Universe	Source counts at large z using HI emission	Deep search f = 100–200 MHz ?	---	Atmospheric Ionospheric No sources at $z > 5$?[a]	Minimum frequency of 300 MHz, 10–50% FC	To z = 4 or 5
	Evolution of objects	Surface brightness (Σ) vs z Angular size (θ) vs z	SENS, RES	$z \sim 3$–5[b]	Confusion (incr. size at larger z?)		
	World models (Closed? Open?)	More accurate studies near $z \sim 1$ of HI from normal galaxies (z vs mag, z vs Σ, θ vs flux)	Each detection in less than a minute	2σ detection 10^{-5}–10^{-6} Jy uniformly distributed across 500 kHz line bandwidth		10–30% FC 3–10% FC	Each detection in less than an hour at z = 1 Each detection in less than an hour at z = 1/2
External galaxies, QSO's, Radio galaxies	Distribution in space/time (see above)	Surveys			50% of such objects are within z = 2[c] Confusion?		
	Evolution	See above					

[a]Reference 20
[b]Reference 21
[c]Reference 22

KEY:
- B = Bandwidth
- DM = Dispersion measure
- FC = Full Cyclops
- ISM = Interstellar medium
- PC = Partial Cyclops
- POL = Polarization
- PSR = Pulsar
- RES = Resolution
- RFI = Radio frequency interference (see complementary document 8)
- SENS = Sensitivity (flux density)
- Σ = Surface brightness
- θ = Angular size
- TBP = Time bandwidth product
- VLA = Very large array
- VLBI = Very long baseline interferometry
- z = Redshift

CHART I.– Continued

Topic	Subhead	Instrumental capabilities or observations needed	FC capabilities	Possible		PC	
				Thresholds	Limits	Dimensions	Capabilities
External galaxies, QSO's, Radio galaxies (cont'd)	Clusters bound or unbound? (mass)	Proper motion	VLBI @ 10^{-4} or 10^{-5} Jy $\approx$400 sec for FC with VLA or Arecibo (10–50 MHz = B)			10–30% FC	VLBI $\approx$ 10–20 min for PC with VLA or Arecibo (10–50 MHz = B)
						1–3% FC	VLBI in 2–4 hr
		Radial velocity	TBP? SENS?				
	Structure (haloes, double sources, bridges, spiral str.)	Mapping, polarization (RES, SENS, TBP)	(~1 kHz res for Faraday Rot.)	VLA-size ($\lesssim$ 1% FC)	Confusion	1–3% FC	30× sensitivity of Westerbork
	HI studies (and mag. field from Zeeman Eff)	SENS, RES, TBP, *circular* POL.				~1 kHz res	Sufficient
	HII studies (recomb. lines) (Faraday Rot.)	B, TBP, *linear* POL, SENS				~1 kHz res	Sufficient
Inter-galactic medium	Deuterium	327 MHz	---			Min freq. 300 MHz	OK
	^{3}He	8.27 GHz	---			Max freq. 10–20 GHz	OK
	Electron densities	From pulsars See below	Time RES, Dispersion See "pulsars"				

CHART I.– Continued

Topic	Subhead	Instrumental capabilities or observations needed	FC capabilities	Possible		PC	
				Thresholds	Limits	Dimensions	Capabilities
Our Galaxy	(Kinematics density, etc.)						
	Small scale dist. of HI, HII; Recomb. lines	~1″–1′ resolution SENS, RES; (B, TBP)	Good	1′ resolution	Confusion	>1% FC	
	Magnetic field Faraday in HII Zeeman in HI in OH?	Linear POL Circular POL		$\Delta\nu$~1 kHz(HI) ~10–100 Hz(OH)	Doppler width?		
	Dynamics from proper motions of stars, parallax	VLBI, SENS (See below)	Sufficient	Sun at 1 kpc		10–50% FC	Sun at 100 pc (not very helpful)?
	Evolution, studies of globular clusters		Good				
Stars	Normal stars to 1 kpc from non-thermal emission	SENS, RES	To 1 kpc	---	Confusion	10–50% FC	To 100–500 pc?
	Astrometry of G stars	VLBI, SENS	Sun at 1–10 pc	*FC*			
	Flare stars Detailed structure M supergiant Transient flares	TIME RES; TBP, B POL, SENS, VLBI? Radiospectroscopy	Radiospectros-copy	8–10 GHz most interesting freq. range?	RFI	→10–20 GHz max. freq.	OK
	1612 and 1720 MHz masers around evolved stars	TBP, SENS, VLBI	Radiospectros-copy				

CHART I.– Continued

Topic	Subhead	Instrumental capabilities or observations needed	FC capabilities	Possible		PC	
				Thresholds	Limits	Dimensions	Capabilities
Stars (cont'd)	Red dwarf flares All T Tauri and P Cygni objects (Stellar winds, formation of planetary systems?)		Throughout Galaxy		RFI	10–50% FC	10–50% of those in Galaxy
Novae, Supernovae, (SN)		SENS, RES	Detect and study all in Galaxy and nearby gal.			10–50% FC	10–50% of all in Galaxy and nearby gal.?
	Time evol. structure (shells, expansion into ISM, early development)	SENS, RES, VLBI? Outriggers	Novae as well mapped as SN are currently	For early development need to be able to rapidly detect in external gal.			See below
	CYG X-3 types		to 50 Mpc			10–50% FC	To ~10 Mpc (10% FC)
	Detection of SN:	SENS	1^m at 60–70 Mpc			1% FC	1^m @ 6–7 Mpc
	Rapid searches in gal. plane			A luxury as SN are relatively rare (1/30 yr)		10% FC	1–2 hr @ 60–70 Mpc
Supernova remnants	Emission mechanism	Circ. POL (to 0.01%)	All SNR's in the local group or 1^m @ 330 kpc			30% FC	Many in the local group 1–2 hr @ 330 kpc

CHART I.– Continued

Topic	Subhead	Instrumental capabilities or observations needed	FC capabilities	Possible		PC	
				Thresholds	Limits	Dimensions	Capabilities
Supernova remnants (cont'd)	Electron dens. (Faraday Rot.)	Linear POL	To Virgo cluster or 1^m det. @ 3–4 Mpc			30% FC	1–2 hr @ 3–4 Mpc
	Spectral index	To 5%	1^m @ 23 Mpc			30% FC	1–2 hr @ 23 Mpc
	Establishment and use of brightness vs size relation as distance scale	SENS, RES (Outriggers?)	To 10 Mpc (further?)			10–50% FC with outriggers	→10 Mpc
	Old remnants and interaction w/ISM	SENS, RES (Mapping)			Confusion ?		
Pulsars	Early history	SENS, rapid searches of nearby gal.	GENERAL: 1. Higher freq. than most present PSR Obs. implies less dispersion by ~100X, but PSR's are only ~10X weaker			10–30% FC Time RES; disp. removal	Detection to ~1–2 Mpc including area increase and dispersion decrease over present situation
	Emission mechanism and nature of coherence	Time resolution Dispersion removal	2. That is, the present dispersion removal techniques are valid over ~100X the distance to presently obs. PSR's (~3 kpc)		RFI		

CHART I.– Continued

Topic	Subhead	Instrumental capabilities or observations needed	FC capabilities	Possible		PC	
				Thresholds	Limits	Dimensions	Capabilities
Pulsars (cont'd)	Use as a means of studying interstellar HI and electron densities on very small scale with or without parallax as independent est. of distance; also proper motions	SENS, POL, Dispersion Removal	GENERAL (cont'd):				
		Complementary HI absorption/ dispersion measurements	3. FC can study all PSR's in Gal. with 5 beams in ~10 yr to present level or better			30% FC, 5 beams	All PSR's in Galaxy to present level in 100 yr
		VLBI for parallax, within a few kpc		1–10% FC			
		VLBI (proper motion)		1–10% FC on grounds of sensitivity		1–10% FC, 5 beams	Throughout Galaxy
	Study intergalactic medium via above except parallax. Obtain better value of mass density in universe		3–30 PSR's in M31 or M33 in few hr (5 beams) Detection in Virgo cluster			30% FC, 5 beams	Top 10 PSR's in M31 in 50 yr
Interstellar medium, molecules, star formation	Electron densities	Compare Pulsar DM with HI absorption or parallax; dispersion removal.					
	HII regions, shells, Recomb. lines	SENS, B, TBP, RES					

CHART I.– Continued

Topic	Subhead	Instrumental capabilities or observations needed	FC capabilities	Possible		PC	
				Thresholds	Limits	Dimensions	Capabilities
Inter-stellar medium, molecules, star formation (cont'd)	Mapping of emission as well as absorption lines in important molecules (HI, OH, CH, H_2CO)	SENS, RES, B B(λ=21,18,9,6cm)→ 1.4–5.0 GHz	10^4 VLA		Confusion for HI	10–30% FC	~10^3 VLA
	Yielding info. on rates and chemistry in clouds						
	Use of *many* weak sources (~1 mJy) for absorption meas. w/great coverage		Detailed coverage @ 1–10″			10% FC	Good coverage at 10″–1′
	Extension to H_2O, NH_3, D, ^3He	Bandwidth 300 MHz–30 GHz					
	Composition studies using undiscovered lines (Mg, Aℓ, Na)	B, TBP, SENS			RFI		
	Studies of clouds and protostellar systems	SENS, RES (→ 1″), 1 mJy	*MAP* individual globules and clouds (~1″–10″ res) and study the very smallest clouds (~1″?) and "clumping" within larger clouds		Confusion	1%–10% FC 10%–30% FC With VLBI or outriggers	*Study* individual globules and clouds (~1′–10′ res) <10″–1′ (resolution within many clouds) Study "clumping" and maybe very small clouds

CHART I.– Continued

Topic	Subhead	Instrumental capabilities or observations needed	FC capabilities	Possible		PC	
				Thresholds	Limits	Dimensions	Capabilities
Inter-stellar medium, molecules, star formation (cont'd)	Maser studies, kinematics	TBP, SENS, VLBI			Doppler broadening ($\Delta\nu$) if masers saturated		
		Circular POL					
	Magnetic field interaction (from Zeeman eff) etc.						
Solar System, Radar + Radio	Surface mapping of terrestrial planets from *Radar*, also surfaces of satellites of Jovian planets, asteroids, comets	Transmitter(s) POL, SENS, RES	2 km resolution at Neptune; SENS α R^{-4} for radar. $\therefore \sim 2$ m resol. for terrestrial planets				
	Similar mapping using radio emission, determination of subsurface temperatures and composition of terrestrial planets at ~500 km res.	SENS, RES				10–30% FC	500 km resolution at favorable encounters (near approach, etc.)
	Probing deep atmospheres of Jovian planets at ~1″–10″ resolution (e.g., zones vs belts)					1–10% FC + outriggers	Map Jovian belts and zones
	Detailed studies of Jovian bursts	TIME RES (B?)	Good			Good	

CHART I.– Concluded

Topic	Subhead	Instrumental capabilities or observations needed	FC capabilities	Possible		PC	
				Thresholds	Limits	Dimensions	Capabilities
Geophysics	Detailed study of tectonic movement (translations *and* rotations of major plates) greatly enhanced by use of many weaker point sources than currently available. Variations in pole of rotation, etc.	VLBI to ~1 mJy		~10–30% FC ? Single antenna (100 m) on each major tectonic plate			

REFERENCES

1. Pauliny-Toth, I. I. K., Kellermann, K. I., Davis, M. M., Fomalont, E. B., and Shaffer, D. B., Astron. J., 77, 265 (1972).

2. Dieter, N., Ap. J., 150, 435 (1967).

3. Reifenstein, E. C., III, Wilson, T. L., Burke, B. F., Mezger, P. G., and Altenhoff, W. J., Astron. & Astrophys., 4, 357 (1970).

4. Zuckerman, B., and Ball, J. A., Ap. J., 190, 35 (1974).

5. Knapp, G. R., Astron. J., 79, 527 (1974a).

6. Knapp, G. R., Astron. J., 79, 541 (1974b).

7. Knapp, G. R., Rose, W. K., and Kerr, F. J., Ap. J., 186, 831 (1973).

8. Knapp, G. R., and Kerr, F. J., Astron. & Astrophys., 35, 361 (1974).

9. Fisher, J. R., and Tully, R. B., Astron. & Astrophys., 44, 151 (1975).

10. Turner, B. E., J. Roy. Astr. Soc. Canada, 64, 221 (1970).

11. Zuckerman, B., and Turner, B. E., Ap. J., 197, 123 (1975).

12. Myers, P. C., and Ho, P. T. P., Ap. J., 202, L25 (1975).

13. Myers, P. C., Ap. J., 198, 331 (1975).

14. Minn, Y. K., and Greenberg, J. M., Astron. & Astrophys., 22, 13 (1973).

15. Höglund, B., and Gordon, M. A., Ap. J., 182, 45 (1973).

16. Simonson, S. C., III. Astron. & Astrophys., 46, 26 (1976).

17. Morris, M., Zuckerman, B., Palmer, P., and Turner, B. E., Ap. J., 186, 501 (1973).

18. Linscott, I. R., Erkes, J. W., and Powell, N. R., Dudley Observatory Report No. 10 (1975).

19. Colgate, S. A., and Noerdlinger, P. D., Ap. J., 165, 509 (1971).

20. von Hoerner, S., in Galactic and Extragalactic Radio Astronomy, Verschuur, C. L., and Kellerman, K. I., eds. (1974) (Springer-Verlag).

21. Refsdal, S., Stabell, R., and deLange, F. G., Mem. R.A.S., 71, Part 3 (1967).

22. Petrosian, V., Third Science Workshop on Interstellar Communication. Held at Ames Research Center, NASA, Moffett Field, Calif. 94035, Sept. 15 and 16, 1975.

Prepared by: Jeffrey N. Cuzzi
SETI Program Office
Ames Research Center

Samuel Gulkis
SETI Project Scientist
Jet Propulsion Laboratory

Contours of equal radio brightness at 21 cm, as measured with the Westerbork Synthesis Radio Telescope (operated by the Netherlands Foundation for Radio Astronomy) of Galaxy M51 compared with the optical image made by the Hale 200-inch telescope. The radio-image provides evidence of the continuity of M51.

6. SETI RELATED SCIENTIFIC AND TECHNOLOGICAL ADVANCES

The material in this section contains specific examples of scientific or technological advances that could arise either as a direct result of a SETI program, or as a result of the science of SETI (see Section II-6).

1. *Observational studies of high brightness contrast systems*: Direct detection of other planetary systems either at visible or infrared wavelengths, will require the development of telescopes capable of studying systems with very high brightness contrast ratios. Particularly valuable contributions could arise from the use of such telescopes to study faint nebulosities associated with quasars and galactic nuclei, for example. One could also examine details hitherto unobservable in short period, mass-transfer binary systems.

2. *Observational studies of small-scale structure in astronomical systems*: The high spatial resolution required of an optical or infrared interferometer used to detect other planetary systems could provide major advances in observational studies of small-scale structure, such as mapping of spiral structure in external galaxies. A high resolution infrared interferometer would prove invaluable for mapping dark interstellar clouds and regions of active star formation.

3. *Improved knowledge of galactic dynamics and distance scales*: The development of an astrometric telescope for the detection of other planetary systems will improve the precision of parallax measurements by two or more orders of magnitude, thereby increasing to several hundred parsecs the range for which parallax studies can provide high precision measures of distance. Use of high sensitivity astrometric telescopes can also provide refined proper motion observations that will allow for more detailed modeling of the dynamics of the Galaxy.

4. *Studies relating to binary frequency*: There are at present few comprehensive studies of the binary frequency of main sequence stars. It would be very useful to carry out such studies over a wide range of spectral types; they could be accomplished with high precision radical velocity machines of the type under development for the task of detecting other planetary systems (see Section II-3). An important aspect of binary studies is the birth function of the secondary bodies. Is it a van Rhijn function, or is it different?

5. *Studies of M-dwarfs*: M-dwarfs constitute the most abundant class of stars, but we know very little concerning their detailed behavior. A coherent program aimed at determining the nature of violent flare events in M-dwarfs would be valuable, not only to a SETI effort, but to stellar astrophysics. At present it is unclear whether flaring is confined to a particular range of spectral type, or whether it is an evolutionary phase through which all M-dwarfs must pass. A flare patrol of M-dwarfs near the galactic pole would be helpful, as those stars are known to be old. Studies of the flare events themselves are important in determining the mechanism, energy spectrum and frequency of these intriguing phenomena.

6. *Studies of the Sun*: The long-term stability of stellar properties, such as luminosity and size, is an important astrophysical topic, and it also relates to a SETI program. The long-term

stability of the Sun's luminosity is currently under debate. The time scale over which luminosity variations might manifest themselves is too long to admit valid conclusions on the question as a consequence of studying one star (e.g., the Sun). However, a proper statistical analysis of a large sample of stars of similar spectral type could answer the question. The classification (spectral and luminosity class) techniques necessary for a stellar census (see Section III-4) will permit high luminosity resolution over a narrow spectral range. High precision radial velocity machines (Section II-3) would make it possible to carry out long-term (months, years) studies of the stability of the Sun's radial velocity (and hence size). This is best accomplished by means of a highly polished sphere in orbit, or mirrors on the ground, which could be used to produce a stellar image of the Sun.

7. *Studies of the luminosity function*: The taking of a stellar census in the solar neighborhood would provide many orders of magnitude more data upon which to model a luminosity function. In conjunction, the high precision astrometric telescopes developed for detection of other planetary systems will both increase the precision with which one can determine the distance to binary star systems, and determine the masses of the stars in a binary system. This will provide a quantum jump in data that can be used to define a mass-luminosity relation over a wide range of spectral type (or mass).

8. *Studies of star and planetary system formation*: It is extremely important to pursue both theoretical and observational studies of star formation. The ability to study fine-structure in dark interstellar clouds (see paragraph 2 above) will provide valuable observational data on the early stages of star formation. It is just as important to carry out theoretical modeling of the star formation process. At present, theory lags observation in this area by a significant amount. Particularly important are studies of angular momentum transport during the dynamic phases of protostellar evolution, studies of the factors controlling the mass spectrum of stars and studies of conditions in circumstellar material. The latter studies provide the much needed link, if such a link exists, between the star formation process and the formation of planetary systems. Theoretical studies of the growth of planets in circumstellar nebulae, and the evolution of such nebulae are also very valuable.

9. *Studies of planetary evolution*: An important aspect of planetary evolution concerns the formation and evolution of planetary atmospheres. Theoretical modeling of this stage of a planet's evolution is fairly rudimentary, and can be improved with additional effort and data. Continued examination of the evolution of the Earth's atmosphere and the atmospheres of other planets in the solar system will provide needed data. Increased studies of the relationship between the Sun and the atmosphere of the Earth in particular, and planets in general, would be very valuable. One would like to be able to estimate how such interactions varied as a function of the spectral type of the parent star in a planetary system.

10. *Studies of chemical evolution and the origin of life*: Great advances have been made in this subject in recent years. However, the transition from chemical evolution to biological evolution (see Section II-1) is not understood at present. Additional laboratory studies in this area, as well as continued attempts to find evidence relating to this transition on the Earth, are very important.

11. *Studies of biological evolution*: The basic mechanism for biological evolution, namely mutations that alter the structure of the DNA molecule in living systems, is understood in a general sense. However, the full panorama of events that might affect or control mutation rates, the tolerance of living systems to mutation rate and to environmental factors is not well understood. Studies in these areas could be of significant value in the field of molecular biology.

12. *Studies of cultural evolution (see Section II-2)*: We are just beginning to appreciate the complexity and diversity involved in the interactions between cultures and individuals within a culture. Studies of social science and the behavior of organisms of various levels of biological complexity are clearly a significant aspect of a SETI program, perhaps most significant following a successful search for extraterrestrial intelligence. Attempts to understand the evolution of, and factors controlling, behavior will not only provide useful input to a SETI program, but would benefit the disciplines of psychology and psychiatry.

13. *Radar astronomy and deep space communications*: Data processing techniques including the strong capabilities in spectrum analysis will be very useful for deep space communications, permitting multiple spacecraft monitoring with small antennas, thus releasing larger aperture facilities from routine tasks. In addition, the application of this technology to monitoring radio frequency interference will greatly assist deep space operations in our crowded spectral environment.

Although no transmitters are presently being considered for any interstellar search system, their inclusion might have enormous advantages with regard to planetary radar astronomy and large distance deep space communications. The great potential of such a system warrants its consideration and further studies should be undertaken.

This listing is not intended to be, nor could it be complete. The intent is to provide some indication of specific areas of scientific research that would provide useful input to a SETI program while also making a major contribution to man's general storehouse of knowledge.

Prepared by: David C. Black
Mark A. Stull
SETI Program Office
Ames Research Center

7. A PRELIMINARY PARAMETRIC ANALYSIS OF SEARCH SYSTEMS

This study was carried out during 1975 and 1976 by the Stanford Research Institute (SRI), Palo Alto, California. The study was concerned with evaluating the comparative cost effectiveness of several different microwave receiving systems that might be used to search for signals from extraterrestrial intelligence. Specific design concepts for such interstellar search systems were analyzed in a parametric fashion to determine whether the optimum location is on Earth, in space, or on the Moon.

System evaluations were performed in terms of a great number of parameters, including the hypothesized number of transmitting civilizations in the Galaxy (N), the number of stars that must be searched to give any desired probability P of receiving one of these signals, the required antenna collecting area, the necessary search time, the maximum search range, and the overall cost. The underlying principle of the method of parametric analysis was that systems are compared only after performance is normalized by requiring that all systems have the same probability of detecting a given signal for any postulated value of N. In practice, this performance normalization was accomplished by requiring that each system examine the same number of stars for any given case (as defined by the values assumed for P and N). Systems with limited sky coverage thus need to search to greater ranges than do systems with more complete sky coverage to examine any given number of stars.

For systems on Earth, primary consideration was given to a Cyclops-type array of conventional parabolic dishes and to an Arecibo-type array of fixed spherical dishes. Preliminary consideration was also given to a Cyclops-type array with a solar power capability. This system has the advantage that costs can be offset by the sale of electricity, but its design must be worked out in detail in a future study. Arecibo-type arrays were found to be slightly more cost effective than Cyclops-type arrays under the assumption that sinkholes are freely available for many dishes to be constructed in the same manner (and for a comparable cost) as the existing dish. This assumption clearly breaks down for large numbers of dishes, and in a future study the possible additional cost of excavating should be evaluated.

For a system in space, a concept was developed that consists of a spherical primary reflector made of lightweight mesh, with three Gregorian subreflectors and feed assemblies that permit three different stars to be searched simultaneously. A disk-shaped shield placed between the antenna and the Earth protects against radio frequency interference (RFI); a series of relay satellites permits the IF signal to be beamed to Earth for processing and to transmit instructions from Earth for station-keeping and attitude control. A number of orbital locations are possible, but for the SRI analysis it was assumed that the system would be at the L3 libration point of the Earth-Moon system, a point at lunar distance but on the opposite side of the Earth from the Moon. Among several desirable features of an orbiting antenna system are the complete sky coverage, the ability to track a given star continuously if a signal is detected, the reduced system temperature, and the potential structural simplicity and longevity made possible by the weightless and benign environment.

For a system on the Moon, primary consideration was given to Cyclops-type arrays and to Arecibo-type arrays. These systems would be located on the far side of the Moon where they would be free from RFI; the systems would be constructed from locally available lunar materials. A lunar colony would have to be established to support the workers needed for construction and for operation and maintenance. The abundance of lunar craters of all sizes makes possible a significant improvement in the design of a lunar Arecibo-type system compared to an Earth-based system of this same type. A major portion of the cost of the actual Arecibo antenna is in the tall towers that support the feed. On the Moon it would be possible to reduce the cost of the antennas substantially by only partially filling a lunar crater with the reflector surface and by suspending both the feed and the dish on long cables extending from the crater rim. For this reason, Arecibo-type systems were found to be significantly cheaper than Cyclops-type systems on the Moon.

Overall cost was defined as the sum of the research and development (R&D) costs, procurement costs, and the operation and maintenance costs over the duration of the necessary search time. The overall cost was found to range from a few hundred million dollars to tens of billions of dollars for any particular system, depending on the values chosen for certain key parameters, particularly the assumed number of transmitting civilizations N, the desired probability of receiving a signal P, and the cost discount factor F. The results of the parametric analysis indicated that systems on the Moon are more expensive than systems on Earth or in space in all cases, even though it is assumed that the cost of transporting material from the Earth to the Moon is only $264/kg. The results also showed that if N is large enough so that the search need be extended only a few hundred light years from Earth, an array of conventional dishes on Earth may be the most cost-effective system (assuming effective protection from radio frequency interference can be obtained). However, for a search that has to extend out to 500 light years or more, for which a minimum of 250,000 stars would have to be examined one-by-one, it was found that there might be a substantial cost and search-time advantage in using a large spherical reflector in space with multiple feeds.

Thus, it was found that there is a reversal in relative cost effectiveness between low and high values of N. Space systems are more expensive than Earth-based systems for large N because of the large investment in R&D that is required before it would be possible to deploy even a moderately sized space system. However, the payoff from this investment increases if larger and larger systems are developed and deployed, so that, for the very large systems that would be required for small values of N, space systems may be cheaper than Earth-based systems. For the particular values of the various system parameters assumed in the SRI study, the overall cost of a system on Earth and of a three-feed system in space would be the same ($\approx$ $11.4 billion) for the case of $N \approx 4 \times 10^5$. A Cyclops-type system for this case would have an antenna collecting area of about 7 km^2, which corresponds to an array of about 890 dishes, each 100 m in diameter. This array could examine about 273,000 candidate stars out to a range of about 535 light years in a search time of about 18 years.

The SRI study was a very preliminary first attempt at comparing costs for alternate locations of an interstellar search system. As such, it should not be regarded as a definitive or in-depth evaluation of interstellar search systems, but as a preliminary treatment in rough order-of-magnitude terms. Many aspects of the cost estimates, especially for systems in space and on the Moon, are necessarily speculative at this time.

Prepared by: Roy Basler
Stanford Research Institute

Space SETI System: Artist's concept of an intermediate size (300-m) space SETI system antenna showing 2 feeds, a relay satellite, RFI shield and a Shuttle type vehicle. Located in geosynchronous orbit or beyond.

8. RADIO FREQUENCY INTERFERENCE

An Interstellar Search System (ISS) is necessarily exceptionally vulnerable to interference by man-made coherent radiations, and the terrestrial radio spectrum is crowded. Therefore an extensive and complex strategy will be required in order to hold interference to an acceptable level. This complementary document discusses the matter only with the intention of demonstrating the seriousness of the problem.

It is convenient to discuss the interference problem in terms of the propagation paths whereby unwanted signals can enter an ISS. The most important paths are:

1. Line-of-sight propagation
2. Reflection from flying or orbiting objects
3. Refractive, diffractive, scatter propagation modes

We first consider line-of-sight propagation. The free-space transmission equation is

$$(P_r/P_t) = \lambda^2 g_r g_t / 16\pi^2 R^2 \tag{1}$$

where

P_r = received power, W
P_t = transmitted power, W
λ = wavelength, m
g_r = receiving antenna gain relative to an isotropic radiator
g_t = transmitting antenna gain relative to an isotropic radiator
R = receiver-transmitter separation, m

Assuming $\lambda = 0.2$ m (or, $\nu = 1.5$ GHz), $g_r = g_t = P_t = 1$, and converting to R, km, we have

$$P_r = 2.53 \times 10^{-10}\ R^{-2}\ \text{(W)}$$

which, on converting to decibel notation, gives

$$P_r(\text{dB}) = -96 - 20 \log R \tag{2}$$

Equation (2) is convenient for this discussion. It represents the received power in dB relative to 1 W, as a function of R. (Should $P_t g_r g_t \neq 1$, add $10 \log (P_t g_r g_t)$ to the right-hand side of eq. (2).)

The lowest received power flux density above which interference will be caused can be calculated from the probable characteristics of an ISS receiving system and is approximately (±10 dBW) independent of the antenna design. The signal-to-noise power ratio (S/N) out of a square-law detector is given by

$$S/N = n(P_r/P_o)^2/(1 + 2P_r/P_o) \quad (3)$$

where

P_r = received-power, W
P_o = total system noise power, W
n = $B\tau$ = number of independent samples that are averaged
B = receiver bandwidth, Hz
τ = averaging time, sec

Setting $(S/N) = 1$, we define the minimum detectable power

$$P_{min} \equiv P_o \frac{(1 + \sqrt{1 + n})}{n}, \quad W \quad (4)$$

Now

$$P_o = kTB \quad (5)$$

where

k = 1.38×10^{-23} JK^{-1}/K^{-1} (where K is the Boltzmann constant)
T = equivalent system noise temperature, K

Combining equations (4) and (5) and inserting reasonable values for the parameters ($T = 5$ K, $B = 0.01$ Hz, $n = 100$ which corresponds to $\tau \cong 10^3$ sec), we have

$$P_{min} = 7.6 \times 10^{-26} \text{ W} = -251 \text{ dBW} \quad (6)$$

Note that P_{min} is at least 50-110 dB less than that common in terrestrial communication services.

Setting $P_r = P_{min}$ as calculated in equation (6), we find

$$R = 5.7 \times 10^7 \text{ km}$$

$$\cong 149 \times \text{the distance to the Moon}$$

That is, using isotropic antennas whose effective cross section is 0.003 m^2, an ISS receiver could detect a 1-W transmitter at this distance. At 2×10^4 km, the same arrangements would produce a receiver output 69 dB above the detection limit; at 200 km, 109 dB.

Clearly, line-of-sight transmissions in and near the ISS reception pass band are a serious matter, whether on the Earth's surface or in distant Earth orbit. Equally, too, out-of-band and harmonic radiations from transmitters in line-of-sight are matters of serious concern.

We can illustrate the problem with respect to reflection from flying or orbiting objects by considering the bi-static radar situation. For simplicity assume the same transmitter and receiver as above, and assume that they are Earth-based and over the horizon from each other. Further, assume there is some object with a scattering (radar) cross section (σ) of 1 m^2, equidistant from the transmitter and the receiver but in line of sight of each. The appropriate relation is

$$P_r = P_t g_t g_r \lambda^2 \sigma / (4\pi)^3 R^4 \times 10^{12} \text{ W} \tag{7}$$

where R is in kilometers. Substituting P_{min} as before, we find R to be 126 km. Further calculation shows that Sky-Lab would return a good signal. In fact, if the experiment were really carried out, Sky-Lab would return many signals to this receiver. And probably many satellites and large pieces of space debris would also do so (see fig. 1).

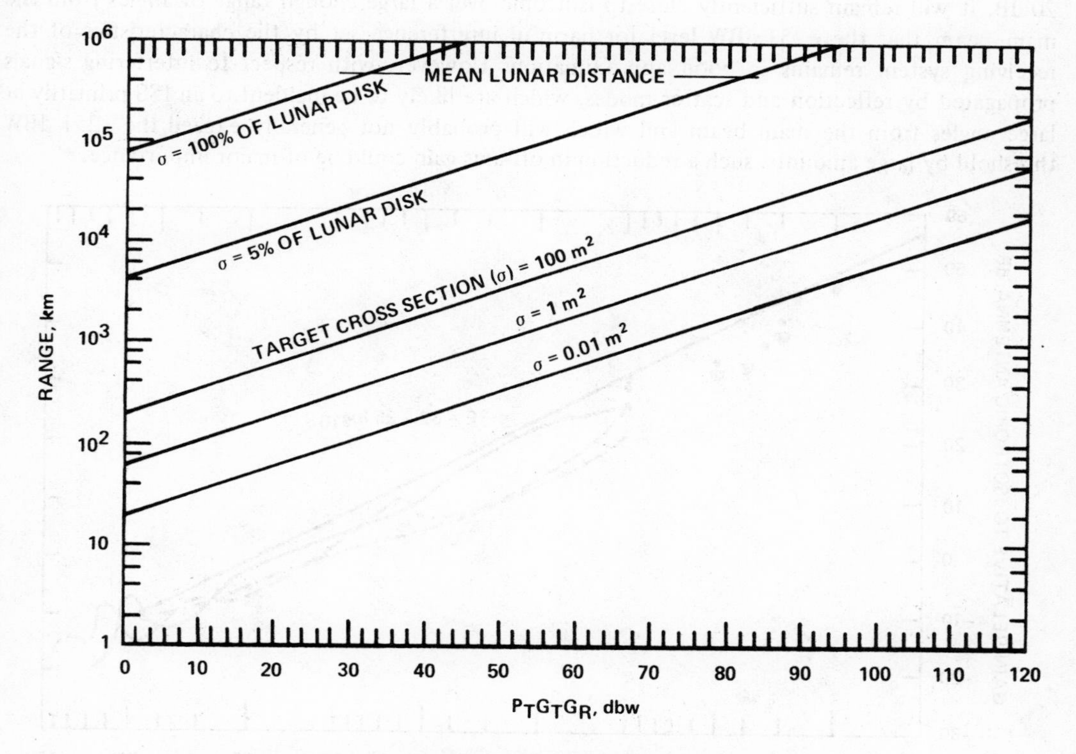

Figure 1.– Bi-static "radar" range for an ISS receiver with $S = N = -238$ dBW at 1.5 GHz.

Discussion of atmospheric-scatter propagation modes is complex. Suffice it to say that an ISS may be able to function in conjunction with properly engineered Earth-bound communication circuits using the same frequencies. In any case, the matter needs careful study.

The nature of the entire directivity pattern of the antennas used in an ISS affects the response of the system to an interfering radiation field. When the physical collecting area of an ISS antenna is greater than some tens of 100-m dishes, the capital and the operating costs of the antennas overwhelm the corresponding electronic and data-handling costs; and a large ISS is costly. Thus there is a strong economic imperative to design for maximum antenna efficiency ($\eta = A$ effective/A geometric). How well one may maximize η while optimizing the antenna response away from the main beam, is a matter not well understood at the present time for lack of theoretical and, in particular, experimental studies. Satisfactory experimental techniques for examining the overall response of an antenna are known. As a result, it is recognized that for a few large antennas with $\eta \cong 0.6$ (which is undesirably low in an ISS context), the off-axis response is down to that of an isotropic antenna at angles from the main beam on the order of 10°-20° when operating at about 1.5 GHz. Only at angles from the main beam of approximately 60° or more does it decrease to more than 10 dB below isotropic (see fig. 2, taken from CCIR Rept. 391-1). Even if novel antenna designs succeed in reducing the off-axis response by an additional 10 to 20 dB, it will remain sufficiently close to isotropic over a large enough range of angles from the main beam that the −251 dBW level for harmful interference set by the characteristics of the receiving system remains in good approximation. However, with respect to interfering signals propagated by reflection and scatter modes, which are likely to be incident to an ISS primarily at large angles from the main beam and which will probably not generally exceed the −251 dBW threshold by large amounts, such a reduction in off-axis gain could be of major importance.

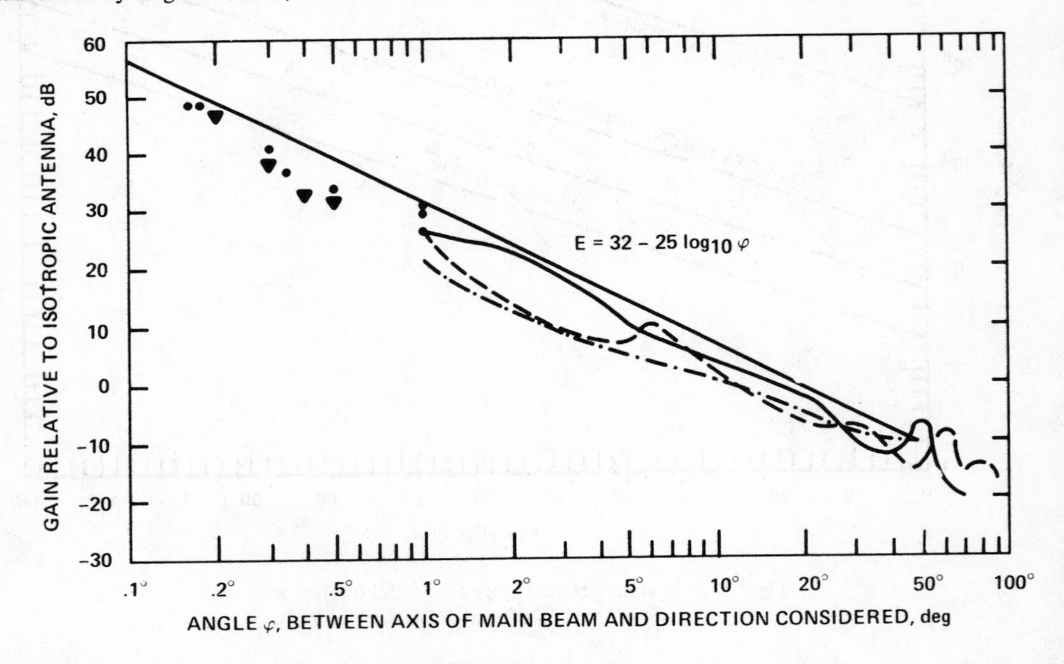

Figure 2.– Peak side lobe levels of radiation patterns for large antennae.

The terrestrial origin of an interfering signal can probably be recognized to a high degree of certainty by a properly operated ISS. But, each detectable interfering signal removes that part of the spectrum from examination for interstellar signals; and strong signals, on the order of −150 dBW at the receiver may totally paralyze an ISS by saturating the masers.

The preceding technical discussion may be summarized as follows. Although improved antenna designs are possible and needed, there is not a hint yet that the $(4\pi - \Omega_{\text{main beam}})$ ISS antenna response can ever be reduced to the range 60-90 dB below the response of an isotropic antenna. Thus operation of an Earth-based ISS is incompatible with transmissions to and from satellites using frequencies in and near the ISS passband. Further studies are needed to determine to what degree the ISS frequencies may be used simultaneously in a variety of services where the signals are kept essentially close to the surface of the Earth. Certainly, an ISS is compatible with a large use of the same frequency band if this use is confined to the surface of the Earth.

The feasibility of constructing a suitable ground-based ISS has been established, and reasonably good cost estimates are possible (see ref. 1). An ISS could also be located in orbit or on the far side of the Moon (see Section III-7). A far-side lunar ISS would need RFI protection only against transmissions originating beyond the lunar orbit; its cost at present, however, appears prohibitive and it probably will not even be possible to build for many years. On the other hand, an orbital ISS might be less costly than an Earth-based system, but it is not clear that cost estimates are yet reliable even to within an order of magnitude. Furthermore, an orbital ISS, unless located far above the geostationary orbit and shielded (an unknown additional cost) is far more vulnerable to RFI than a ground-based system. Indeed, an unshielded ISS would require a total clearing of the observation band (see Section III-9) because all Earth-based transmitters in the hemisphere facing it, as well as satellite transmitters, would be line of sight. Of course a shielded ISS in orbit would require only about the same protection as an ISS on the far side of the Moon.

Thus, whatever the ultimate ISS location, some kind of RFI protection seems desirable. Furthermore, it is likely that preliminary Earth-based searches with existing large radio telescopes, as well as perhaps a small ISS (say, five to ten 100-m dishes) will be conducted for at least a decade prior to the start of operations with larger systems, both to search for strong signals and to perfect techniques, receivers, and data processors. These preliminary searches will require essentially the same degree of protection as a full-scale Earth-based system, because that is dictated by the value of P_{min} (eq. 6) and not the antenna collecting area. Thus the development of a SETI program in the reasonably near future depends upon reaching an accommodation with the way in which the electromagnetic spectrum is to be used. Fortunately, the use of the 1400 to 1727 MHz band is not yet fully developed and a move starting now to accommodate ISS requirements would forestall large investments now in the planning and conceptual stages.

In 1979 there will be a general World Administrative Radio Conference (WARC). Such meetings are held by the member countries of the International Telecommunications Union (ITU) approximately every 20 years to determine radio spectrum allocations among the various competing users. It is the purpose of the upcoming WARC to review, thoroughly, existing allocations across the entire spectrum and revise them in accordance with the envisaged needs of the remainder of the century. Thus if an ISS is to be built to operate in the 1400 to 1727 MHz band (or in any other band), it will be necessary to obtain formal international protection for it in 1979.

At present, satellite use of the 1400 to 1727 MHz band is relatively small; however, the MARISAT system is currently being brought into operation there, and the AEROSAT and NAVSTAR systems will probably use the band starting at some time in the 1980's; and the FCC has long-range plans to open the band to television broadcasting. Beyond this, it can be anticipated that plans for other satellite systems will be developed, and that as a result, an enormous equipment investment will be made in the 1980's and 1990's, both in the 1400 to 1727 MHz band and elsewhere in the spectrum, up to about 30 GHz. The effect of this huge prospective investment – perhaps tens of billions of dollars – will be to create great economic pressure against any attempt at a subsequent WARC to revise whatever allocations are devised in 1979. Certainly any protection given the 1400 to 1727 MHz band for SETI after 1979 will result in much greater economic dislocation.

On the other hand, much present use of the 1400 to 1727 MHz band is compatible with the operation of an ISS, or could be if properly engineered. In contrast to bands below 1400 MHz and above 1727 MHz, where one finds powerful aircraft acquisition radars with EIRP ranging from 10^7 to more than 10^9 W, terrestrial transmitters in this band are primarily low-power. Satellite users are still relatively few, and satellite equipment is still designed to be amortized generally in about five years, although construction for much longer lifetimes is planned for the 1980's and thereafter. The necessity of replacing equipment as it becomes obsolete provides an opportunity to engineer the new equipment to operate at a different frequency without incurring excessive costs. Then, too, a large fraction of present equipment can be cheaply modified for operation in a neighboring or other appropriate band. Since a large ISS would probably not be operational before 1985 at the earliest and large-scale preliminary searches will not be made before the early 1980's, satellite transmissions in the band can be tolerated until then, and perhaps longer if need be, to avoid economic hardships. Thus, if prompt action is taken, a plan that will protect the 1400 to 1727 MHz band with minimum economic impact can be developed and implemented. This plan would be in compliance with United States telecommunications law, which requires that, in the allocation of frequencies, public interest considerations, such as those that motivate a SETI program, prevail over economic ones; it should be based on the resolution of the Science Workshop (see Section I, Conclusion 2, p. 11, and Section III-9).

REFERENCES

1. Oliver, B. M.; and Billingham, J.: Project Cyclops, A Design Study of a System for Detecting Extraterrestrial Intelligent Life. NASA CR114445, 1972.

Prepared by: Mark A. Stull
Charles L. Seeger
SETI Program Office
Ames Research Center

9. PROTECTION OF A PREFERRED RADIO FREQUENCY BAND

The Science Workshop, especially at its second, third, and fourth meetings, debated all considerations concerning whether the probability of finding signals from an extraterrestrial civilization is maximized in any particular frequency band. To the extent this problem requires knowledge of the motivation of such civilizations, it cannot be solved, but there are physical and philosophical arguments which imply that the frequency band between 1400 and 1727 MHz should have high priority for a search effort (see Section II-4). Because of the sensitivity of any interstellar search system (ISS), it is very important that the only telecommunications services which operate in this band be those that will not cause harmful interference to an ISS. In 1979 a general World Administrative Radio Conference (WARC) will be held; it will allocate world-wide use of the radio spectrum and allocations made then are likely to determine spectrum usage for the remainder of this century. The Science Workshop recognized the importance of obtaining protection of the 1400 to 1727 MHz band for SETI use at the 1979 WARC and, to emphasize this need, adopted the following resolution.

STATEMENT ON THE REQUIREMENTS FOR PROTECTION OF AN INTERSTELLAR SEARCH SYSTEM (ISS) FROM RADIO-FREQUENCY INTERFERENCE

In recognition of the rapidly advancing national preparation for the 1979 general World Administrative Radio Conference (WARC), the Science Workshop adopts the following final statement of policy:

1. There are important frequency bands for a search for radio signals from extraterrestrial intelligent civilizations. These are:

   a. 1.400 to 1.427 GHz
   b. 1.427 to 1.727 GHz

The 1.400 to 1.427-GHz band is of interest because interstellar transmissions may take place around the hydrogen line, while the 1.427 to 1.727-GHz band is located between the hydrogen and hydroxyl lines and lies near the minimum of the noise background. 1.400 to 1.427 GHz is currently allocated exclusively to the radio astronomy service and may be shared with it, while 1.427 to 1.727 GHz may be shared with services whose use will not cause harmful interference to the operation of an ISS.

2. Existing radio telescopes are already being used to search for radio signals from extraterrestrial civilizations, while the feasibility of constructing a very large ground-based ISS has been established. The performance of any ground-based instrument will, however, be seriously degraded by radio-frequency interference, primarily from line-of-sight transmitters. The only identified alternatives to an Earth-based ISS are:

- A space-based ISS, and
- An ISS sited on the far side of the Moon

Both of these are possible in the future, but we do not know at what cost. Furthermore, a space-based ISS, unless shielded at additional expense, remains vulnerable to interference from satellite and ground-based transmitters; while an ISS on the far side of the Moon is vulnerable to all transmissions originating beyond the lunar orbit. Thus, there exists a need for RFI protection. We strongly recommend:

a. That the U.S. undertake immediate studies to determine detailed frequency protection requirements for an ISS, and submit the results of such studies to the 1977 Final Meeting of the International Radio Consultative Committee (CCIR) for inclusion in the supporting documents of the 1979 WARC, and

b. That the U.S. prepare and present to other administrations at the 1979 WARC a proposal which will include

i) Allocations for new satellite systems at frequencies outside the protected bands.

ii) Appropriate frequency sharing criteria for uses compatible with the operation of an ISS.

iii) Technical criteria for allowable spurious radiation from out-of-band uses.

iv) Phase-out of interfering uses now operating in the protected bands.

Subsequent to the Science Workshop deliberations and during the preparation of the final report, the SETI radio frequency protection need received its first international recognition. Reproduced below is Addendum No. 1 to Volume II, XIIIth Plenary Assembly of the CCIR, Geneva, 1974.

Note by the Director, C.C.I.R.

Subsequent to the publication of Volume II (Space Research and Radio Astronomy) of the documents of the XIIIth Plenary Assembly of the C.C.I.R., a new text relating to search for extraterrestrial life has been submitted for adoption by correspondence, in conformity with the provisions of No. 308 of the International Telecommunication Convention, Torremolinos, 1973.

It has received more than the twenty approvals necessary for its adoption by the Members of the I.T.U. and has therefore become an official Question of the C.C.I.R.

The text is as follows:

QUESTION 17/2

RADIOCOMMUNICATION REQUIREMENTS FOR SYSTEMS TO SEARCH FOR EXTRATERRESTRIAL LIFE

(1976)

The C.C.I.R.,

CONSIDERING

(a) that many scientists believe intelligent life to be common in our galaxy;

(b) that electromagnetic waves are presently the only practical means of detecting the existence of intelligent extraterrestrial life;

(c) that it is believed to be technically possible to receive radio signals from extraterrestrial civilizations;

(d) that, although it is not possible to know the characteristics nor to predict the time or duration of these signals in advance, it is reasonable to believe that artificial signals will be recognizable;

(e) that, while an artificial radio signal of extraterrestrial origin may be transmitted at any frequency, it is technologically impractical to search the entire radio spectrum but the band searched should be sufficiently wide to make detection of a signal reasonably probable;

(f) that technological and natural factors which are dependent on frequency determine our ability to receive weak radio signals;

(g) that while the search for radio signals from extraterrestrial civilizations has already begun, more sensitive systems will be in use by the 1980's which could receive harmful interference from very weak man-made signals;

(h) that it is necessary to share the bands in which the search is conducted with other Services;

(j) that aváilable technology will allow a search for these signals from the earth, from earth-orbit, and, eventually, from the moon and to minimize interference, certain locations on earth and in space may be preferred;

DECIDES that the following question should be studied:

1. what are the most probable characteristics of radio signals which might be broadcast by extraterrestrial civilizations and the technical characteristics and requirements of a system to search for them;
2. what are the preferred frequency bands to be searched and the criteria from which they are determined;
3. what protection is necessary for receiving systems conducting a search for artificial radio signals of extraterrestrial origin;
4. what criteria will make operation of a search system feasible in shared, adjacent and harmonically related bands of other Services;
5. what is the optimum search method;
6. what are the preferred locations, on earth and in space, for a search system?

Addendum No. 1 to Volume II, XIIIth P.A. of the C.C.I.R., Geneva, 1974

Prepared by: Mark A. Stull
SETI Program Office
Ames Research Center

10. RESPONSES TO A QUESTIONNAIRE SENT TO LEADING RADIO OBSERVATORIES

At the conclusion of the second full meeting of the Workshop, it was agreed to survey the directors of representative radio observatories on a number of matters where the apparent direction of SETI and the concerns of those observatories overlapped strongly, especially in the utility of wide-band simultaneous reception.

The following letter was sent to about 35 such directors from the published lists. Seventeen answers were received from organizations in the United States and in three other countries. The responses to the questions are summarized below; the identity of the respondents is filed in the records of the Workshop. One foreign response included an original contribution to the problem; several other responses included reprints and other valuable information.

Overall, we learned that there was not much unpublished work in SETI, and that most radio observatories could foresee some forms of cooperation between their ordinary scientific investigations and work on behalf of SETI.

MASSACHUSETTS INSTITUTE OF TECHNOLOGY

DEPARTMENT OF PHYSICS

CAMBRIDGE, MASSACHUSETTS 02139

Room 6-308

August 29, 1975

Dear

About fifteen years ago it first became clear that the normal development of the science of radio and radar astronomy had had an unexpected consequence: we humans found ourselves in actual possession of the means of signalling to an assumed counterpart installation across interstellar distances. To this day no other communications technique is capable of such a reach.

That fact stimulated the proposal to listen - though not to transmit - for such signals from elsewhere. Whatever your estimate of the probability that there exist distant counterparts, possibly much advanced ones, of our technology, it is plain that eventual empirical tests of the conjecture imply interests parallel to the broader concerns of most radio astronomers. I write to draw upon that convergence of interest, to ask your help in hope of mutual advantage.

A scientific working group has been established lately in the USA to examine and to report upon the question of a search for interstellar communications. I am currently Chairman of the group - their names are listed below - which includes participants from a wide range of disciplines, astronomers with a wide variety of specialty, information and communications experts, biologists, experienced engineers.

A somewhat similar effort to ours has lately been reported by a Board of the Academy of Sciences of the USSR (1).

It is premature to announce any decision on optimum channels and procedures, or on the scale of effort which makes sense. But we regard the frequency range from about 1 Ghz to a few tens of Ghz as a major candidate. That forms the basis of our mutual interest.

Within this form of search, two or three distinct modes can be seen. The initial Soviet emphasis seems to be devoted to a search with wide-band non-directive means for short pulses, looking to time coincidences at distant receivers (2) to distinguish the signature of an extraterrestrial source. The opposite mode is fully as attractive: seeking a narrow-band signal with highly directive antenna beams, much improving the signal-noise ratio at the price of a search in direction and frequency (3,4). There is a kind of intermediate: a relatively wide beam is directed at an external galaxy which fills it, hoping to search all at once a large number of potential stellar sources, all of them of course very remote (5). There are other possibilities (6).

While pioneer searches have been published, it is clear that the task has been barely begun. We want to ask you and your staff to help us appraise the feasibility of a mutual approach, the use of existing large antennas (with low system temperatures) to establish limits on such signals as a by-product of present scientific research. Narrow-band signals would much reward the use of multichannel spectrum analyzers for simultaneous search over some 10^6 up to 10^9 channels, each of the order of 1 Hz width, in the band from 1.4 to 1.7 Ghz. Carefully chosen target directions would be examined in a plan which included a certain amount of area search in unexpected directions as well.

We ask you to respond to a brief list of questions, whose tendency is clear. Please give us the opinion of your group, for only the present expert users of such instruments can supply realistic answers. It goes without saying that we would interested in any other expression you wish to send us, beyond the response, brief or more extended, you might wish to make to our questions.

We should appreciate hearing your thoughts by around 15 October 1975. If we can expand or clarify our request in any way, please do not hesitate to ask.

Sincerely,

Philip Morrison

Philip Morrison
Institute Professor

PM/mb
references on separate page

Membership of the Scientific Working Group on Interstellar Communications:

Bracewell, Ronald
Brown, Harrison
Cameron, A. G. W.
Drake, Frank
Greenstein, Jesse
Haddock, Fred
Herbig, George
Kantrowitz, Arthur
Kellermann, Kenneth
Lederberg, Joshua
Lewis, John S.
Morrison, Philip (Ch.)
Murray, Bruce C.
Oliver, Bernard M.
Sagan, Carl
Townes, Charles H.

REFERENCES

1) See the report: The CETI Program, by the Scientific Council on the Radio-Astronomy Problem Area, Academy of Sciences of the USSR. Astron. Zh., 51, 1125-32 (September-October 1974). English translation in Sov. Astron., 18, 669-675 (March-April 1975), American Institute of Physics.

2) The first account of such a pulse search appears in: Troitsky, V.S., pp. 259-60 in the volume CETI, C. Sagan, editor (full citation below). A larger system is proposed in (1).

3) One report of a US effort at 1.4 GHz (since the classical Ozma work of 1960): Verschuur, G.L., Icarus 19, 329-40 (1973).

4) A Soviet narrow-band search (near 0.9 Ghz): Troitsky, V.S., Starodubtser, A.M., Gershtein, L.I. and Rakhlin, V.L., Astron. Zh. 48, 645 (1971); English translation in Sov. Astro. 15, 508 (1971). One study is proceeding at the Algonquin Radio Observatory in Canada (Bridle and Feldman) using 22 Ghz.

5) A first search was made at the Arecibo Radio Observatory in June, 1975, using a multi-channel analyzer and a beam directed at a number of nearby galaxies, including M31 (Drake and Sagan, unpublished).

6) An exploratory search was made from the orbiting UV satellite Copernicus for UV laser lines from three nearby stars (Herbert Wishnia, 1974).

We mention two recent general references for orientation:

i) A semi-popular brief up-to-date review, Drake, F. and Sagan, C., Scientific American 232, 80 (May, 1975).

ii) The report of a small international conference held in September, 1971 at the Byurakan Astrophysical Observatory, Armenian SSR. It is published as the volume: Communication with Extraterrestrial Intelligence (CETI), edited by C. Sagan. The MIT Press, Cambridge, Massachusetts and London, England (1973).

QUESTIONS FOR USERS OF RADIO TELESCOPES

1) What is the stability of your local oscillators?
   a) in typical use
   b) the best you have available

2) Do you now have or are you now planning to obtain a multi-channel spectrum capability (well beyond the usual widths of 0.1 to 10 kHz)?

3) Could astronomical research benefit from simultaneous multi-channel analyzers with 10^6 to 10^9 channels of 1 Hz width? Comments?

4) Comment on the present state of the art of such systems. What characteristics would you require? How would you proceed to achieve them? What directions of development seem to you to offer most promise?

5) If such an analyzer were made available to you would you consider using it, via an IF tap or with spare receivers, to "hitch a ride" for coherent signal search during normal observing time?

6) Have you ever engaged in any search for coherent or "intelligent" signals at your facilities? If so, please estimate the total hours used. Have the results been reported? Was the search a by-product of other observations? Explain.

Please respond by 15 October 1975 to:

Philip Morrison
Room 6-308
M.I.T.
Cambridge, Mass. 02139

Responses... organized by question

Question 1)

What is the stability of your local oscillators?
a) in typical use
b) the best you have available

Respondent:

#1 a) 10^{-9}/day; b) 10^{-10}/day
2 use NRAO and Haystack equipment
3 1 in 10^{10}
4 10^{9}
5 ±1 in 10^{9}/min; ±1 in 10^{8}/wk.
6 a) 10^{-9}; b) 10^{-12}
7 1 in 10^{6}/day
8 1 in 10^{11}
9 a) 10^{-8}; b) 10^{-9}
10 ≈ 10^{-11}/month
11 1 in 10^{12}
12 ----
13 3 phased arrays at 74 MHz
14 ∿10^{-8}
15 1 in 10^{10}
16 10^{-12}
17 a) ±50kHz; b) 1 kHz

Question 2)

Do you now have or are you now planning to obtain a multi-channel spectrum capability (well beyond the usual widths of 0.1 to 10 kHz)?

Respondent:

#1 Have 64 channel autocorrelator
2,3,4,5 No
6 Now: 100 channels with 1kHz, 2kHz, 10kHz or 30kHz
Planned: 1000 channels with 100kHz or 200kHz
7,8,9 No
10 Now: autocorrelator ∿100Hz
No plans for spectrometer with a higher frequency.
11 Now: multi-channel spectrometer with a 100-channel one-bit autocorrelator
Planned: 1024 channel autocorrelator.

Question 2) Continued

12 ----
13,14 No
15 Have 32 channel spectrum analyzer and 1024 channel autocorrelator.
16 Planned: 1024 autocorrelator with minimum channel width of 200Hz
17 Planned: 50 channel 10kHz bandwidth filter.

Question 3)

Could astronomical research benefit from simultaneous multi-channel analyzers with 10^6 to 10^9 channels of 1Hz width? Comments?

Respondent:

#1,2,3,
4,5 Yes
6 For most spectroscopy 1kHz resolution is sufficient; the more channels the better.
7 10^6-yes
10^9-don't know
8 ----
9 ----
10 Yes
11 10^6-10^9 channels would overresolve spectral features
12 ----
13 Yes
14 Do not forsee uses for such high frequency
15 Possible, but expensive
16 No astronomical need for frequency resolution $\lesssim$ 100Hz.
17 100Hz seems to be smallest bandwidth useful for studies of known natural processes.

Question 4)

Comment on the present state of the art of such systems. What characteristics would you require? How would you proceed to achieve them? What directions of development seem to you to offer most promise?

Question 4) Continued

Respondent

#1,2 ----
3 Feasible, using autocorrelators
4 ----
5 Achieve very narrow resolution through use of autocorrelation spectrometer.
6 Autocorrelators or multifilters for 1-100mhz total coverage. Acousto-optical or multifilter techniques for larger coverage.
7,8,9 ----
10 Increase the usefulness of such a spectrometer if it could be split into several parallel receivers.
11 Optical data most promising approach.
12 ----
13 Working continuously on fine time resolution from weak signals after removing interstellar dispersion.
14 Autocorrelator techniques to achieve such large capacity.
15 Problem with 10^6-10^9 multi channel analyzers is the drift of the center of frequency of each channel; shouldn't exceed $\sim$0.1 Hz.
16 For VLA spectral-line back end, a system is contemplated which would have $\sim 2\times10^5$ channels divided among the 351 interferometer pairs.
17 2 approaches
 a) coherent optical processing method
 b) Fast Fourier Transform algorithm.

Question 5)

If such an analyzer were made available to you would you consider using it, via an IF tap or with spare receivers, to "hitch a ride" for coherent signal search during normal observing time?

Respondent:

#1,2,3 Yes
4 To some extent
5,6 Yes
7 ----
8 Best to use antennas with large collecting area.
9 Possibly
10,11,12 Yes
13 Concerning technical improvements in pulsar observations: versatile very fast digital processing unit to perform the dispersion removal over wide bandwidths.
14 Yes

Question 5) Continued

15	Combined answer with question 4).
16,17	Yes

Question 6)

Have you ever engaged in any search for coherent or "intelligent" signals at your facilities? If so, please estimate the total hours used. Have the results been reported? Was the search a by-product of other observations? Explain.

Respondent:

#1,2	No
3	Yes
4	Not seriously
5	Yes; no scientific purpose
6	See attachment
7,8,9,10,11	No
12	3 searches: a) OZMA/Drake b) Verschuur c) Palmer and Zuckerman (unpublished)
13	No
14	Search for electromagneitc pulses from the galactic center.
15	Yes
16	No
17	Yes; searching for narrow-band radiation near the hydrogen line.

Prepared by: Philip Morrison
Professor of Physics
Massachusetts Institute of Technology

11. THE SOVIET CETI REPORT

Researchers in the USSR have long been interested in the detection of radio signals originating from extraterrestrial intelligence. The Soviet have named their program CETI or Communication with Extraterrestrial Intelligence. The acronym SETI (Search for Extraterrestrial Intelligence) was adopted by the Workshop and by the Ames Research Center to differentiate our own efforts from those of the Soviet Union and to emphasize the *search* aspects of the proposed program.

The Soviet plans for their CETI efforts have been summarized in "The CETI Program," Sov. Astron., vol. 18, no. 5, March-April 1975, which has been reprinted here, in total, with permission of the American Institute of Physics, whose cooperation is gratefully acknowledged. This article was translated from "Scientific Council on the Radio-Astronomy Problem Area, Academy of Sciences of the USSR, Astron., Zh., *51*, 1125-1132 (September-October 1974)."

The CETI program

Scientific Council on the Radio-Astronomy Problem Area, Academy of Sciences of the USSR Astron., Zh., 51, 1125-1132 (September-October 1974)

In March 1974 the Board of the Scientific Council on the Radio Astronomy Problem Area, Academy of Sciences of the USSR, considered and approved a Research Program on the Problem of Communication with Extraterrestrial Civilizations. The Program was developed by the Search for Cosmic Signals of Artificial Origin section of the Council on Radio Astronomy, on the basis of recommendations made at the Soviet National Conference on the Problem of Communication with Extraterrestrial Civilizations held at the Byurakan Astrophysical Observatory in Armenia in May 1964, and the Soviet-American CETI conference held at Byurakan in September 1971. The projected program was reported to the 7th Soviet National Conference on Radio Astronomy, which convened at Gor'kii in 1972.

The program as outlined below is here published in detail with minor abridgments.

RESEARCH PROGRAM ON THE
PROBLEM OF COMMUNICATION
WITH EXTRATERRESTRIAL CIVILIZATIONS

PART I. INTRODUCTION

1. Formulation of the Problem
2. Principal Fields of Research on the Problem of Extraterrestrial Civilizations
3. Principles for Developing a CETI Program
4. Organizing Arrangements

PART II. SEARCH FOR COSMIC SIGNALS OF ARTIFICIAL ORIGIN

1. Search for Sources and Selection by Preliminary Criteria

   1.1. Radio Surveys of the Sky
   1.2. Selection of Sources by Angular Size and Investigation of Their Spatial Structure
   1.3. Investigation of Selected Galactic and Extragalactic Objects
   1.4. Search for Signals from Stars in the Immediate Solar Neighborhood
   1.5. Search for Signals from Galaxies in the Local Group
   1.6. Search for Signals with a Detection System Covering the Entire Sky
   1.7. Search for Probes
   1.8. Measurement of Cosmic Background Radiation in the Wavelength Range $20\,\mu - 1$ mm
   1.9. Investigation of Absorption and Phase Transparency of the Interstellar Medium in the Range $20\,\mu - 1$ cm
   1.10. Sky Surveys in the Range $10\,\mu - 1$ mm
   1.11. Search for Infrared Excesses in Stars of Suitable Spectral Type

2. Investigation of the Radiation Structure of the Selected Objects and Methods of Analysis for Identifying Sources Suspected of Being Artificial

3. Instrumentation Projects for Seeking Radio Signals from Extraterrestrial Civilizations

PART III. DECODING OF SIGNALS

PART IV. CONCLUSION

PART I

INTRODUCTION

1. FORMULATION OF THE PROBLEM

Over the past few years the scientific community has begun to show increasing interest in the problem of contacts with extraterrestrial civilizations.

The question of whether intelligent life exists elsewhere in the universe in some form or another has been posed in every era throughout the development of science. Yet it is only now, thanks to major advances in astronomy, biology, cybernetics, information theory, radiophysics and radio engineering, and the conquest of space, that it has become possible for the first time to progress from purely speculative arguments on this subject to systematic scientific investigation. The achievements of modern science have led to a deeper understanding of the fundamental aspects of the problem. The hope of establishing communication with extraterrestrial civilizations today rests on a scientific basis. This endeavor has come to be called the CETI problem (Communication with Extraterrestrial Intelligence).

Advances in radiophysics and radio engineering have played a decisive role here. Modern radio techniques would enable signals transmitted across interstellar distances to be detected and recorded. Thus it is already feasible to plan research and experiments for detecting signals from extraterrestrial civilizations. From the very outset such investigations can and should rest on the achievements of radio astronomy, in which much experience has now been gained in identifying and analyzing sources of cosmic radio emission. Progress in infrared and optical astronomy will also make a vital contribution, in conjunction with rapidly advancing developments in laser technology. Also of great value for the practical organization of a CETI program will be the achievements of information theory and other branches of cybernetics, which offer general methods for studying problems of information transfer, as well as such areas of mathematics as game theory, the theory of tactics, and searching theory. With these disciplines as a basis a special CETI strategy can be devised. At the present time, then, the technical ways and means are available for practical steps to be taken in the CETI field.

The possibilities of radio communication with extraterrestrial civilizations were first analyzed by G. Cocconi and P. Morrison (United States, 1959), who showed that under certain conditions signals could be received from extraterrestrial civilizations with the radio technology available at that time. In 1960 the first practical steps were taken in the United States to look for such signals at 21-cm wavelength (F. D. Drake, Project Ozma). At present a more comprehensive signal search program is under way at the U.S. National Radio Astronomy Observatory (Project Ozma 2), and a Project Cyclops has been studied[1] by Stanford University in collaboration with the Hewlett-Packard Company; it would ultimately cost tens of billions of dollars. In the Soviet Union, work comparable to Project Ozma is being carried out at the Gor'kii Radiophysics Institute.

These projects presuppose that extraterrestrial civilizations have a level of technological development analogous to our own. In this event one would anticipate discovery of monochromatic radiation similar to the radiation from ordinary transmitters on the earth.

In 1964 another signal search concept was put forward in the Soviet Union, whereby extraterrestrial civilizations would be expected to have a very high level of development. Several radio-astronomical criteria for an artificial source were formulated on the basis of this concept. Experiments were subsequently devised at the Shternberg Astronomical Institute of Moscow University and at the Institute for Space Research, Academy of Sciences of the USSR, for the purpose of examining a number of peculiar sources in order to see whether they might satisfy the proposed criteria.

Searches for pulsed signals from space have recently been undertaken in the Soviet Union at the Gor'kii Radiophysics Institute, the Institute for Space Research, and the Shternberg Astronomical Institute. Various aspects of the CETI problem are being studied individually not only at these institutions, but by the Council on Radio Astronomy of the Academy of Sciences of the USSR, the Moscow Power Institute, the All-Union Electrical Engineering Correspondence Institute of Communications, the Russian Language Institute, Academy of Sciences of the USSR, and elsewhere.

One other approach, entailing a search for signs of astro-engineering activity by highly developed extraterrestrial civilizations (in particular, from surveys of infrared radiation), has been suggested by F. J. Dyson (U.S.).

In 1960, R. N. Bracewell (U.S.) made the important suggestion that a search ought to be conducted for radio signals from space probes, which even now might conceivably be present in the solar system.

The U.S. National Academy of Sciences held a special conference in 1962 on the problem of communication with extraterrestrial civilizations.

Major contributions toward formulating and discussing the CETI problem have been made by the Soviet National Conference on the Problem of Communication with Extraterrestrial Civilizations[2] (Byurakan, 1964) and the Soviet-American CETI conference[3] (Byurakan, 1971). Topics pertaining to the CETI problem[4] have been considered at various other conferences and meetings both in the Soviet Union and abroad.

The present program has been developed from recommendations by the 1964 Soviet National Conference, the 1971 International Conference, and the 7th Soviet National Conference on Radio Astronomy in 1972.

2. PRINCIPAL FIELDS OF RESEARCH ON THE PROBLEM OF EXTRATERRESTRIAL CIVILIZATIONS

The problem of extraterrestrial civilizations comprises an intricate complex of topics in philosophy and sociology as well as natural science. Within the domain of this broad interdisciplinary problem a narrower area is to be considered: the CETI problem. This represents a separate task confronting science and technology, including theoretical and experimental work on searching for extraterrestrial civilizations, as well as modeling the basic links in the CETI system. But a successful result will depend on resolving a number of fundamental questions that form the heart of the extraterrestrial-civilization problem.

It is convenient to distinguish two groups of problems for planning the investigations.

Group A. Fundamental Problems of Extraterrestrial Civilizations Involving Communication

1. *Astronomical matters*. Cosmogony. Discovery of planets, planetlike bodies, and congealed stars. Sky surveys conducted in various parts of the electromagnetic spectrum. Examination of some peculiar sources. Investigation of organic compounds in cosmic objects.

2. *Life*. A more precise definition of the concept of "life." Possible existence of nonprotein life forms. Origin of life on the earth; possible alternative origins of life on other cosmic bodies, and in interplanetary and interstellar space. Exobiology. Laws of biological evolution and their exobiological generalization.

3. *Intelligence and intelligent systems*. Refinement of the concept of "intelligence" or "reasoning." Models of an intelligent system. Theory of complex self-organizing systems. Information contacts in complex systems. Symbolic systems; language. Problems in the theories of knowledge and reflection; construction of models.

4. *Mankind*. Analysis of the laws governing the development of civilization on the earth. Special characteristics of the rise and development of different civilizations worldwide. Forecasting. Development and mastery of the space environment.

5. *Information transfer*. Optimum methods of communicating information.

These topics are being dealt with independently of the CETI problem itself and therefore are not considered in the present program (except for the sky surveys).

Group B. Problems Pertaining Directly to CETI

1. *Aspects of the theory of cosmic civilizations*.

2. *Contacts between cosmic civilizations; possible types of contact and their consequences.*

3. *Modes of intercourse between cosmic civilizations.* Linguistic media to be devised for establishing information contact between "intelligent" systems.

4. *Procedures and scientific-technological basis for seeking signals from extraterrestrial civilizations.* Development of signal search techniques. Influence of the cosmic medium on exchange of signals between civilizations. Choice of optimum electromagnetic wavelength range. Criteria for identifying signals from extraterrestrial civilizations. Characteristics of "call letters." Design of search instrumentation. Modeling of individual links in the CETI system. Computer modeling.

5. *Searches for signals from extraterrestrial civilizations.*

6. *Deciphering of signals.*

7. *Searches for astro-engineering activity of extraterrestrial civilizations.* Although the main emphasis in this program is given to efforts to find signals in the radio range and to the development of suitable techniques and equipment, a more complete program should also include planning with regard to other aspects of the CETI problem.

3. PRINCIPLES FOR DEVELOPING A CETI PROGRAM

The present program has been drawn up on the basis of the following initial propositions.

1. Efforts to detect extraterrestrial civilizations should proceed smoothly and systematically, and should extend over a prolonged period of time. The program is oriented in this direction from the very outset. It would be a great mistake to build a program in contemplation of rapid and easy success.

2. Investigations should be based on a specially devised program (or group of programs) which would be revised and perfected as time passes. The program should provide every opportunity to take advantage of existing technology (radio telescopes, antenna systems and associated instrumentation), and should also envision the development of specialized techniques and equipment for coping with the CETI problems.

3. The program will recognize that astrophysical information will be acquired as a byproduct of the search for signals from extraterrestrial civilizations. When actual investigations are undertaken it will be necessary to analyze carefully the question of what astrophysical applications can be pursued during search activities.

4. In view of the uncertainty in our a priori knowledge as to the character of signals from extraterrestrial civilizations, the program should entail parallel studies in several directions.

4. ORGANIZING ARRANGEMENTS

1. Matters relating directly to CETI (Group B) are currently being worked on in a random fashion; such research is not being planned properly.

But the problem has now reached a state requiring more earnest organizational efforts. If the research is to proceed successfully one cannot avoid creating a number of organizations and institutions to deal with appropriate branches of the problem; these should be fully staffed and furnished with equipment and materials. The design of search instrumentation and detection systems calls for organizational enterprise on an industrial scale. Henceforth all scientific and engineering work in the field of searches for signals from extraterrestrial civilizations ought to be placed under the guidance of a pilot organization.

2. Matters pertaining to Group A topics are being pursued independently of the CETI problem, and often quite separately from one another. In the very near future thought should be given to ways in which these studies can be coordinated and purposefully integrated into the CETI plane.

PART II

SEARCH FOR COSMIC SIGNALS OF ARTIFICIAL ORIGIN

Initial Premises

The program for seeking signals from other civilizations rests on the two assumptions mentioned above as to their level of development:

1. A level of technological development and, in particular, the technology of radio communications comparable to that on the earth.

2. A level of technological development and modes of communication far more advanced than our own.

For the first case, from considerations of power capability, signals should be sought primarily from nearby stars. The character of the signals might be analogous to the signals of transmitters on the earth; that is, narrow-band signals may be anticipated (it would then be possible to generate special call signals distributed over a wide frequency range so as to facilitate searches with respect to frequency).

In the second case the detection problem would be considerably simpler because of the much greater power capacity. However, uncertainty might arise both in the coordinates of the sourc (as such civilizations would not necessarily be associated with stars) and in the character of the signal. Presumably the signals of highly developed extraterrestrial civilizations would most likely be wide-band. Although this circumstance would ease the search in frequency, it would require the formulation of definite criteria enabling such signals to be distinguished from wide-band radiation of natural origin. Regions that might be worth searching could be located near the nuclei of our own and other galaxies, or associated with certain peculiar sources.

This program presupposes the need for parallel research at both the levels 1 and 2 above, with no a priori assumptions as to the character of the signals. Thus the problem of methodically searching for signals from extraterrestrial civilizations will include the following aspects: a) a search for sources of radiation with respect to direction, frequency, and time; b) an analysis of the structural properties of the radiation; c) determination of its artificial nature.

Accordingly, this program provides for the following steps in research activity:

1. Search for and selection of sources according to preliminary criteria.

2. Examination of the radiation structure of the sources selected.

3. Analysis of the results obtained, in an effort to identify artificial sources.

1. SEARCH FOR SOURCES AND SELECTION BY PRELIMINARY CRITERIA

This step entails both conducting sky surveys in the optimum wavelength range from the CETI standpoint followed by source selection according to preliminary criteria, and analysis of sources already known. The result should be the compilation of a catalog of objects promising from the CETI standpoint and warranting further study in more detail.

As criteria for preliminary source selection, the following may be used and are adopted in this program: small angular size; distinctive spatial structure of the source; distinctive spectrum (special behavior of the energy distribution, presence of narrow-band features, special shapes in the spectrum such as rectangular features, and so on); unusual character of time variability; distinctive polarization properties (such as a regular alternation of left- and right-circular polarization in the spectrum). As astrophysics progresses and new data emerge, this list could be refined or revised.

1.1. Radio Surveys of the Sky

Frequencies. Since a CETI signal might be restricted in its spectrum, in order to discriminate promising sources the surveys should fully cover the entire shortwave portion of the radio-astronomy range (frequencies of 1-100 GHz), which according to present indications is the most advantageous region for CETI purposes.

In the initial phase surveys could be made at discrete frequencies for which appropriate instrumentation is already available or is under development.

Anetnnas. The principal antennas for conducting surveys in the USSR could be the RATAN-600 (the periscopic part) in the wavelength range 0.8-2.1 cm, and the antenna for the millimeter range being developed at the Gor'kii Radiophysics Institute.

Instrumentation. Radio receiving equipment for the surveys should include:

1. Continuum radiometers with $\Delta f \approx 0.1f$ and $T_n \approx 100°K$, equipped with facilities for magnetic-tape recording and with auxiliary output devices for the detection of information-carrying signals according to anticipated criteria.

2. Radiometric systems with as wide a band as possible ($\Delta f \approx 0.5f$).

3. Spectral radiometers for surveys at the frequencies of individual radio lines, such as $\lambda\lambda$ 21, 18, and 1.35 cm.

1.2. Selection of Sources by Angular Size and Investigation of Their Spatial Structure

a. Preliminary selection of sources with an angular size smaller than the antenna beam, and investigation of their spatial structure. No specialized equipment would be required.

b. Measurement of the angular (as well as the linear) size and spatial structure of sources by the interstellar-scintillation technique. This method affords the highest angular resolution. However, for surveys at wavelengths shorter than 10 cm or in the case of very close sources ($R < 10$ pc) the interferometer method is best. The radio telescopes may either by ground-based or stationed in space, permitting earth–earth, earth–space, and space–space baselines.

1.3. Investigation of Selected Galactic and Extragalactic Objects

On the basis of astrophysical evidence presently available, studies of the following objects would be of interest from the CETI standpoint: globular clusters, representing the oldest objects in our galaxy; the galactic center, a region containing 10^9 stars; the galaxies of the Local Group; certain nearby radio galaxies and quasars.

The aim of such investigations would be to discover anomalies in the radio emission of these objects from the point of view of the criteria given above.

Instrumentation. Antennas with an effective area greater than 1000 m^2 should be used for most of the objects. It is recommended that the size, spatial structure, and radio variability be investigated by the scintillation technique with earth–earth, earth–space, and space–space baselines, using a radio telescope in space and others on the ground. Spectroscopic and polarimetric properties could be examined with the same equipment and with the RATAN-600 radio telescope.

This type of work is closely linked with the central problems of radio astonomy and can be carried out in the course of conventional radio-astronomical research.

1.4. Search for Signals from Stars in the Immediate Solar Neighborhood

It is proposed that individual nonvariable stars of suitable spectral type be observed. Initially the observations may be confined to monitoring all appropriate stars to a distance of 10-100 light years from the sun; eventually out to 1000 light years.

If the monitoring extends over a period of several years, then in order to inspect all suitable stars within a radius of 100 light years with a single antenna operating continuously, the total observing time for each star would be about one hour, or no more than a few hours.

Antennas. For this survey it would be desirable to use several radio telescopes with an effective area of $\approx 1000\ m^2$. Observations could begin with smaller-sized telescopes.

The primary task in the initial phase of investigation would be to detect radio emission from stars, because the intrinsic radio emission of solar-type stars is weak. As for the kind of signals to be expected, it would be advisable to begin by searching for very simple signals – pulsed, monochromatic, and the like. This program will impose corresponding requirements on the instrumentation.

In the future one might hope to search for signals by sending automated space probes to the nearest stars.

1.5. Search for Signals from Galaxies in the Local Group

Any search for signals from galaxies would enable an enormous number of stars (roughly 10^{10}-10^{11}) to be covered simultaneously. The nearest galaxies are of interest to us inasmuch as we might be of interest to them.

Since the number of objects is small, a continuous monitoring service could be organized, with observations extending over several years. The optimum wavelength range for each object can be made more definite by taking its background into consideration.

Antennas. For a continuous service, weakly directional antennas would be used with a beam covering the galaxy being observed ($\theta \approx 1$-$3°$). In addition, it would be desirable to employ larger antennas with an effective area of $\approx 1000\ m^2$ so that parts of galaxies could be investigated and repeated surveys made of each galaxy.

Instrumentation would be similar to that used in the search for signals from stars.

Special emphasis should be given to a search for pulsed signals (if they have a low averaged power not influencing the observed flux density of the galaxy, they could still have a high peak power, ensuring that $P_s/P_n > 1$) as well as monochromatic signals.

In studying galaxies with the aim of detecting signals from extraterrestrial civilizations, one will no doubt acquire information of astrophysical character pertaining to those galaxies.

1.6. Search for Signals with a Detection System Covering the Entire Sky

The detection system would here be designed for continual monitoring of the radiation of the whole sky in the optimum anticipated wavelength range. Such a system would enable transient sporadic signals coming from any direction to be recorded.

Regular monitoring of the entire sky would require a network of stations located at different points on the globe or in space.

For stations placed on the earth, the number needed is determined by the condition that objects be simultaneously visible from at least two widely spaced stations (in order to discriminate local interference), so that at least four would be required altogether.

If stations are placed in space at a large distance from the earth, just two would be sufficient. The best plan would evidently be to put these stations in orbit around the moon, because when they pass behind the moon maximum suppression of terrestrial interference would be assured. One very promising location for a space station would be the Lagrangian point in the earth–moon system located beyond the moon.

Building a system of large directional radio telescopes to cover the whole sky would be a major undertaking financially. Thus at the outset we would propose that investigations be organized with nearly omnidirectional antennas that would record only the strongest signals. Subsequently the stations would be equipped with high-efficiency antennas, and the detection capability of the system could be raised gradually.

Frequencies. In due course the entire shortwave part of the radio-astronomy range should be examined. At present the search should be limited to discrete frequencies, using equipment now available or under development.

The main effort at first should be concentrated toward searches for sporadic pulsed signals. It would be advisable to begin this program by utilizing equipment that satisfies the following requirements:

The recording system should detect pulses lasting from 1 sec to 10^{-8} sec (with a choice of time constants differing by decade factors).

The equipment should be provided with a system for obtaining the dispersion measure for an approximate distance estimate of a source, and for discriminating terrestrial interference.

Observations should be controlled by the Time Service to a precision adequate for these measurements. A rough determination of the direction toward a source could be made from the lag in the signals at different stations.

Restrictions on the observing stations. The principal requirement is a low noise level. Such stations will generally be located in regions difficult of access. Perhaps a full suppression of interference and elimination of the effects of the earth's atmosphere, as well as a more accurate determination of positions, will demand that some of the stations (at least two) be placed in space at a large distance from the earth.

1.7. Search for Probes

The possible discovery of probes sent from extraterrestrial civilizations and now located in the solar system or even in orbit around the earth warrants particular attention. To search for these rapidly moving objects the system of constant monitoring of the whole sky should be supplemented by specially designed radio direction-finding systems. Initially it would be possible to use existing installations intended for space communications and radar observations.

1.8. Measurement of Cosmic Background radiation in the Wavelength Range 20 μ – 1 mm

Such measurements are needed to identify regions of minimum intensity and establish more accurately the optimum wavelength range for CETI purposes.

1.9. Investigation of Absorption and Phase Transparency of the Interstellar Medium in the Range 20 μ – 1 mm

1.10. Sky Surveys in the Range 10 μ – 1 mm

These surveys should be made in an effort to find objects associated with the engineering activity of extraterrestrial civilizations.

It is proposed that the sky be surveyed in this wavelength range and that the spectra of any objects detected be investigated.

1.11. Search for Infrared Excesses in Stars of Suitable Spectral Type

The purpose of this search would be to detect thermal radiation inherently emanating from large-scale works of engineering that may have been constructed in circumstellar space.

Telescopes and equipment for surveys 1.10 and 1.11. A special mountain observatory should be built (so as to diminish losses of infrared radiation due to absorption by atmospheric water vapor) with a telescope about 2.5-3 m in diameter, and a specialized satellite carrying analogous instrumentation should be launched.

The receivers would be high-sensitivity bolometers with a selection of filters.

2. INVESTIGATION OF THE RADIATION STRUCTURE OF THE SELECTED OBJECTS AND METHODS OF ANALYSIS FOR IDENTIFYING SOURCES SUSPECTED OF BEING ARTIFICIAL

Each of the objects selected by the preliminary criteria will be investigated more carefully, and the research program will be modified in each instance depending on the results of the measurements obtained during the preliminary selection process.

It would be desirable to conduct the analysis of cosmic radio waves along the following lines: examination of the shape of the radio spectrum as a whole for envelopes and carrier waves; inspection of the fine structure of the spectrum; recordings of rapid variability of the carrier and envelope spectra; establishment of the carrier and envelope distribution functions; determination of how the polarization (especially circular polarization) depends on the carrier frequency and on time; investigation of recordings obtained by analog and digital detection of carriers and envelopes at different frequencies and bands for both.

The data obtained in this manner would be used to identify sources suspected of being artificial. The methods and criteria for identification should include an analysis of all data from the standpoint both of astrophysics (comparison with known and possible natural astrophysical objects) and cybernetics and information theory (comparison of the statistical properties and structure of the signal with known or anticipated types of communication).

3. INSTRUMENTATION PROJECTS FOR SEEKING RADIO SIGNALS FROM EXTRATERRESTRIAL CIVILIZATIONS

To ensure that the avenues of research described above will be carried out, two instrumentation projects, CETI 1 and CETI 2, are proposed, as follows.

CETI 1 Project (1975-1985)

1. A ground-based system continuously monitoring the entire sky, comprising eight stations with nondirectional antennas supported by detection equipment capable of covering the whole optimum wavelength range.

2. A satellite system continuously monitoring the entire sky, comprising two space stations with nondirectional antennas and fully covering the optimum wavelength range.

3. A system of low-directivity antennas of 1-3° beamwidth for a continuous survey of nearby galaxies (subsection 1.5). These antennas might conveniently be located at the same stations where the sky is continuously monitored by nondirectional antennas.

CETI 2 Project (1980-1990)

1. A satellite system continuously monitoring the entire sky and equipped with antennas of large effective area.

2. A system of two widely spaced stations having large (effective area $\approx 1\ \text{km}^2$) semi-rotatable antennas for synchronized reception, searches for signals from specific objects, and analysis of selected sources.

These instrumentation complexes could be used not only for CETI work but for a variety of important astrophysical problems.

In addition, individual parts of the program could be carried out with other radio telescopes in conjunction with the plans of radio-astronomy institutions (sky surveys, investigations of peculiar sources, and so on).

PART III

DECODING OF SIGNALS

One of the most important problems in need of solution for CETI purposes is to work out deciphering techniques specifically applicable to extraterrestrial communications (in the absence of any a priori information as to the language, method of encoding, and character os the signals). Within the present program a leading role should be assigned to logically formal deciphering techniques, comprising algorithms that can be implemented only by computer, and enabling a given linguistic entity to be designated according to a maximum of special "estimator" functions computed from counts made on the test being analyzed. The decoding procedure for signals from extraterrestrial civilizations that is to be developed as part of this program may be broken down into several steps.

1. *Preliminary analysis of signals.* In this step the alphabet of the elementary signals (messages) would be established.

2. *Determination of type of text language organization.* Three types of organization are presumably possible: pictorial (image transmission), linguistic (analogous to the structure of languages on the earth), and formalized (as with logical computer languages or algorithms).

3. *Disclosure of grammatical system of the language.* The properties of the test explained by the internal structure of the language would be ascertained at this stage, that is, the parametrization of the frames and alphabetic gradations used to express model languages. The grammar of a humanoid language would be determined, or for formalized systems, the axioms and rules of construction and derivation.

4. *Disclosure of the semantics of texts under investigation.*

5. *Development of methods for translating the decoded language into familiar languages.* This step would, in particular, encompass techniques for automatic compilation of bilingual dictionaries and structural correspondence lexicons.

The development of signal decoding techniques is closely related to research on image recognition, automatic classification and encoding (for example, some of the algorithms worked out in the decipherment analysis might be applied to study the structure of branches of the national economy), and work in the area of computer translation and automated abstracting.

PART IV

CONCLUSION

Even in their initial stage investigations of the CETI problem can be of important cognitive and applied value, and can serve as a source of useful information and a stimulus in many fields of science and technology.

This working program, which includes experimental and theoretical projects of immediate concern, is provisionally directed toward the next 10-15 years. It should form part of a more complete program taking the long view, and it also offers a basis for developing specialized, detailed programs in particular areas of research that fall within the scope of the CETI problem.

Continual revision and improvement should be made in the program as data are accumulated in all fields pertaining to CETI and as individual search programs are conducted.

REFERENCES

[1] Project Cyclops: A Design Study of a System for Detecting Extraterrestrial intelligent Life (NASA-CR-114445), Stanford Univ. and NASA Ames Res. Cen., Moffett Field, Calif. (1972).

[2] G. M. Tovmasyan, ed., Extraterrestrial Civilizations (Proc. Byurkan conf., May 1964), Armenian Acad. Sci. Press (1965) [Israel Program Sci. Transl., No. 1823 (1967)].

[3] C. Sagan, ed., Communication with Extraterrestrial Intelligence (CETI) (Proc. Byurakan conf., Sept. 1971), MIT Press (1973).

[4] S. A. Kaplan, ed., Extraterrestrial Civilizations: Problems of Interstellar Communication, Nauka, Moscow (1969) [NASA TT F-631 (1971)].

12. SEARCHES TO DATE

In addition to on-going radioastronomical observations, there have been several deliberate attempts to detect signals of extraterrestrial intelligent origin, all with negative results. In most cases, only a few select objects have been observed, at a few discrete frequencies using relatively wide bandwidths and at only moderate to low sensitivities. Although most laudable, these observations require enormous equivalent isotropic radiated power on the part of other civilizations to be detectable.

INVESTIGATOR	OBSERVATORY	DATE	FREQUENCY OR WAVELENGTH	TARGETS
DRAKE	N.R.A.O.	1960	1,420 MHz	EPSILON ERIDANI TAU CETI
TROITSKY	GORKY	1968	21 AND 30 CM	12 NEARBY SUNLIKE STARS
VERSCHUUR	N.R.A.O.	1972	1,420 MHz	10 NEARBY STARS
TROITSKY	EURASIAN NETWORK, GORKY	1972 TO PRESENT	16, 30 AND 50 CM	PULSED SIGNALS FROM ENTIRE SKY
ZUCKERMAN PALMER	N.R.A.O.	1972 TO PRESENT	1,420 MHz	~600 NEARBY SUNLIKE STARS
KARDASHEV	EURASIAN NETWORK, I.C.R.	1972 TO PRESENT	SEVERAL	PULSED SIGNALS FROM ENTIRE SKY
BRIDLE FELDMAN	A.R.O.	1974 TO PRESENT	22.2 GHz	SEVERAL NEARBY STARS
DRAKE SAGAN	ARECIBO	1975 (IN PROGRESS)	1,420, 1,653, AND 2,380 MHz	SEVERAL NEARBY GALAXIES
DIXON	O.S.U.	1973 TO PRESENT	1420 MHz	AREA SEARCH

13. THE MAINTENANCE OF ARCHIVES

Rational search by humankind through a hyperspace of the size and complexity contemplated here requires systematic compaction and storage of the results as they come in. Uncommon care should be exercised in deciding what data to discard. In the near future, a plan should be prepared for a relatively long-lasting storage procedure, one providing easy access to the stored data, not only to colleagues in the search, but also to others who have an interest for any variety of reasons. Its construction will require careful thought, some experience, and recognition that this data bank has enormous growth potential. Probably, such a plan would itself grow through a number of stages in response to comments following dissemination of a preliminary plan.

It is attractive to suppose some appropriate journal would publish the plan and its development, and perhaps continue with periodic reports of the results to date in summary form, with clear information on how to gain the full details. The initial publication should be co-authored by representatives of several of the groups in the field. The Byurakan-founded international CETI committee could easily serve to invite such attention.

The magnitude of the permanent data storage problem presented by an extensive, state-of-the-art SETI installation is worth illustration. Consider just the preposed receiver – 300 MHz instantaneous bandwidth; 1 Hz frequency resolution (bin width); 6 polarization channels; and 16 bits per output bin per second; 100 sec per target direction (for 100 unit observations). Thus there will be approximately 3×10^{12} bits to be disposed of every 100 sec. A large fraction of this torrent of data is redundant, since it is just plain noise. Sorting and compacting routines are an economic necessity for all recognizable spectral structures 2 or 3 sigma above the mean noise level. Then, the remainder of the data must be discarded.

What should be saved for future reference? Here is a partial list:

1. Housekeeping data – when, by whom, with what installation; circumstances of the observation; essential system parameters, etc.

2. Identified (interfering) coherent signals – general characteristics (frequency, drift rates, bandwidth, etc.).

3. Unidentified coherent signals – general characteristics and estimated nature.

4. Astrophysical data – continuum radiation as a function of frequency; a wealth of (primarily) interstellar line profiles; variable or intermittent radiations (pulsars, etc.); etc.

5. Pattern recognition and compaction algorithms in operation.

6. Comments by observers.

And all this should be available to interested parties for an indefinitely long time, easily, and without requiring highly specialized equipment.

As a final remark, we stress the need to favor storage over discard, but with good judgment. Any clear hint of structure above the calm sea of background noise may be the clue.

Prepared by: Charles L. Seeger
SETI Program Office
Ames Research Center

14. SELECTED ANNOTATED BIBLIOGRAPHY

The most detailed bibliography of interstellar communication published to date is that of Mallove and Forward. Those readers wishing to explore the subject further will find the bibliography in reference 1.

1. Mallove, E. F., and Forward, R. L. Bibliography of Interstellar Travel and Communication. I. J. of Brit. Interplanetary Soc., *27*, 921–943 (1974); II. J.B.I.S. *28*, 191–219 (1975); III. J.B.I.S. *28*, 405–434, 1975).

2. Oparin, A. I. Life: its nature, origin and development. Oliver and Boyd, London (1962).
   A summary of the writings of the distinguished Soviet biochemist, including some of the earliest theories of chemical evolution, first proposed by the author in the 1920's.

3. Haldane, J. B. S. The Origins of Life, in New Biology, No. 16. Penguin, London (1954).
   Further original writing on the origin of life and chemical evolution, written before the critical experiments of Miller were carried out.

4. Miller, S. L. and Orgel, L. E. The Origins of Life on Earth. Prentice Hall (1974).
   A summary of twenty years of laboratory and field work in chemical evolution, written by two of the leading investigators in this field.

5. Lederberg, J. Exobiology. Approaches to Life Beyond the Earth. Science *132*, 393–400 (1960).
   The distinguished American geneticist explores all aspects of life outside the earth, or exobiology. Included are suggestions for searching for life in our solar system.

6. Cocconi, G., and Morrison, P. Searching for Interstellar Communications. Nature *184*, 844 (1959).
   The original paper in which proposals are made for the detection of signals of extraterrestrial origin at frequencies near the hydrogen line at 21 cm.

7. Cameron, A. G. W. (Ed.) Interstellar Communication. Benjamin Press (1963).
   The first collection of original essays on interstellar communication, with stimulating papers by Morrison, Oliver, Bracewell, Cameron, Shklovskii, Townes, Huang and Van Hoerner. Included also are two papers by Frank Drake, one describing the first definitive attempt to detect signals – Project Ozma.

8. Shklovskii, I. S., and Sagan, C. Intelligent Life in the Universe. Holden-Day, New York (1966).
   The first substantive book on all aspects of interstellar communication. It is a collective venture between two distinguished U.S. and Soviet authors. Now a classic.

9. Kaplan, S. D. (Ed.) Extraterrestrial Civilizations. Problems of Interstellar Communication. NASA TTF-63 (1971).

A collection of Soviet papers from a number of years previously, covering many aspects of interstellar communication. In English.

10. Oliver, B. M., and Billingham, J. Project Cyclops. A Design Study of a System for Detecting Extraterrestrial Intelligent Life. NASA CR114445 (1972).

The first detailed conceptual design study for a system for detecting signals from extraterrestrial civilizations. Project Cyclops is a terrestrial system of phased radiotelescopes with a highly sophisticated data processing system.

11. Kreifeldt, J. G. A formulation for the number of communicative civilizations in the Galaxy. Icarus, *14*, 419–430 (1971).

The first statistical evaluation of the possibilities for interstellar communication over galactic time scales.

12. Sagan, C. (Ed.) Communication with Extraterrestrial Intelligence. M.I.T. Press, Cambridge, Massachusetts (1973).

A detailed and exciting account of the first international meeting on communication with extraterrestrial intelligence, held at Byurakan, Armenia in 1971 under the auspices of the U.S. and Soviet Academies of Science.

13. Oliver, B. M. State of the art in the detection of extraterrestrial intelligent signals. Astronautica Acta, *18*, 431–439. Press (1973).

A summary of possible engineering systems for the detection of signals from extraterrestrial intelligent life, including the main features of Project Cyclops.

14. Ponnamperuma, C., and Cameron, A. G. W. (Eds.) Interstellar Communication. Scientific Perspectives. Houghton Mifflin, New York (1974).

A collection of essays on interstellar communication presented in a lecture series at the NASA Ames Research Center in 1971. A very readable successor to Cameron's original book.

15. Sagan, C., and Drake, F. The Search for Extraterrestrial Intelligence. Scientific American *232*, 80–89 (1975).

A condensed account of activities in interstellar communication to date, including a description of those attempts which have already been made to listen for signals.

16. Oliver, B. M. Proximity of Galactic Civilizations. Icarus, *25*, 360–367 (1975).

The author examines, on a statistical basis, the number of times in galactic history that two civilizations will have emerged close to each other and at the same time, perhaps stimulating each to search for other civilizations.

17. Bracewell, R. N. The Galactic Club Intelligent Life in Outer Space. W. H. Freeman and Company (1974).

A highly readable synopsis of some of the major unanswered questions about the nature and distribution of intelligent life, including suggestions for contact by interstellar probes.

15. INTERSTELLAR COMMUNICATIONS SCIENCE WORKSHOP MEMBERS

Dr. Philip Morrison, Institute Professor and Professor of Physics at the Massachusetts Institute of Technology, is Chairman of the Science Workshops on Interstellar Communication. Professor Morrison is both a distinguished theoretical physicist and a respected scholar-philosopher whose ecumenical interests embrace the broad sweep of human and scientific history from the origin of the universe to the origins and definitions of intelligent life itself. He was one of the first scientists to predict that knowledge as to whether life exists on other planets may not be beyond our reach.

Dr. Ronald N. Bracewell, Professor of Electrical Engineering, Stanford University, has made numerous contributions in the field of radioastronomy and has been interested in the existence of extraterrestrial intelligence for many years. He is the author of the book "The Galactic Club: Intelligent Life in Outer Space."

Dr. Harrison S. Brown, Professor of Geochemistry, California Institute of Technology, has made extensive contributions to science in the study of transuranium elements, meteorites, geochronology, physics and chemistry of the solar system, science and public policy, population problems, environmental problems, and natural resources.

Dr. A. G. W. Cameron is Professor of Astronomy at the Harvard College Observatory, Harvard University, and Associate Director for Planetary Sciences at the Center for Astrophysics. His areas of research interest include nucleosynthesis and associated areas of nuclear physics, stellar evolution, supernova explosions, neutron stars, quasars, physics of the interstellar medium, origin and development of the solar system, and physics of planets and planetary atmospheres. He is co-editor of "Interstellar Communication: Scientific Perspectives," a collection of papers on the search for extraterrestrial intelligence.

Dr. Frank D. Drake is Professor of Astronomy at Cornell University and the Director of the National Astronomy and Ionosphere Center, Arecibo, Puerto Rico. His numerous contributions to the field of radio and radar astronomy are well known and widely acclaimed. He conducted the first organized search for extraterrestrial intelligent radio signals called Project OZMA.

Dr. Jesse L. Greenstein, Professor of Astrophysics, California Institute of Technology, has made extensive contributions in the study of the interstellar medium and stellar evolution. He was the Chairman of the Planetary Detection Workshops.

Dr. Fred T. Haddock, Professor of Astronomy, University of Michigan, developed the radioastronomy facility at Michigan and is presently the Director of the University of Michigan Radioastronomy Observatory. Prof. Haddock is active in both ground as well as space based radioastronomical observations.

Dr. George H. Herbig, is Professor of Astronomy, University of California, Santa Cruz. His research specialties include spectra of variable and peculiar stars, optical absorption spectroscopy of interstellar material and the early evolution of stars.

Dr. Arthur Kantrowitz is Senior Vice-President and Director of AVCO Corporation and Chairman and Chief Executive Officer of the AVCO Everett Research Laboratory. He first became well known for his research in physical gas dynamics, and particularly for his pioneering application of the shock tube to high temperature gas problems.

Dr. Kenneth I. Kellerman is Staff Scientist, National Radio Astronomy Observatory, Charlottesville, Virginia. Dr. Kellerman is also the Chairman of the National Advisory Committee, Owens Valley Radio Observatory.

Dr. Joshua Lederberg is Professor of Genetics and Biology, and Chairman of the Department of Genetics at the Stanford University School of Medicine. In 1958, he was awarded the Nobel Prize in medicine for studies on the organization of the genetic material in bacteria. Dr. Lederberg was the Chairman of the Cultural Evolution Workshop.

Dr. John S. Lewis is Associate Professor of Chemistry, Earth and Planetary Sciences, Massachusetts Institute of Technology. He has made numerous contributions in the study of the composition, structure and origin of planetary atmospheres and the application of thermodynamics to problems of composition and origin of meteorites.

Dr. Bruce C. Murray is Professor of Planetary Science at the California Institute of Technology and has made extensive contributions in the field of planetary imaging. In April, 1976 he became Director of the California Institute of Technology's Jet Propulsion Laboratory.

Dr. Bernard M. Oliver is Vice-President of Research and Development for the Hewlett-Packard Corporation. Dr. Oliver was Co-Director of the 1971 Stanford/Ames Research Center Summer Faculty Fellowship Program for the Design Study of a System for Detecting Extraterrestrial Intelligent Life, Project Cyclops.

Dr. Carl Sagan is Director, Laboratory for Planetary Studies, and Professor of Astronomy and Space Sciences at Cornell University, where he is also Associate Director for the Center for Radio Physics and Space Research. His principal research activities are in the physics and chemistry of planetary atmospheres and surfaces, in space vehicle exploration of the planets, and on the origin of life on Earth.

Dr. Charles H. Townes, University Professor at the University of California since 1967, received the Nobel Prize for his role in the invention of the maser and laser. Internationally known for his research on the interaction of electromagnetic waves and matter, and also as teacher and government advisor, he is affiliated with the Department of Physics on the Berkeley campus, and engaged in research in astrophysics.

PART-TIME CONSULTANTS

H. R. Brockett	National Scientific Laboratories
Eugene Epstein	Aerospace Corporation
Robert Machol	Northwestern University
Marcia Smith	Congressional Research Service

NASA people who have been associated with the Workshop activities

NASA Headquarters
Joseph P. Allen
William E. Brunk
Hugh Fosque
William Gilbreath
John Naugle
Ichtiaque Rasool
Nancy G. Roman
Stanley Sadin
Robert E. Smylie
Karlheinz Thom
Richard S. Young

Jet Propulsion Laboratory
Donald Davies
Robert Edelson
Richard Goldstein
Samuel Gulkis
Alan Hibbs
Michael Janssen
Gerald Levy
Robert Powell
Donald Rea
Nicholas Renzetti
Anthony Spear

Ames Research Center
John Billingham
David Black
Vera Buescher
Alan Chambers
Mary Connors
Jeff Cuzzi
Donald De Vincenzi
Richard Johnson
Harold Klein
Hans Mark
Ray Reynolds
Charles Seeger
Joe Sharp
Mark Stull
John Wolfe

Goddard Space Flight Center
Thomas Clark
Robert Cooper
George Pieper

Johnson Space Center
Yoji Kondo

Marshall Space Flight Center
Ernst Stuhlinger